EXTRA EXERCISE BOOK
for

Moore's
The Basic Practice of Statistics
Second Edition

David S. Moore
Darryl K. Nester

W. H. Freeman and Company
New York

ISBN: 0-7167-4338-8

Copyright © 2001 by W. H. Freeman and Company

No part of this book may be reproduced by any mechanical, photographic, or electronic process, or in the form of a phonographic recording, nor may it be stored in a retrieval system, transmitted, or otherwise copied for public or private use, without written permission from the publisher.

Printed in the United States of America

First printing 2001

CONTENTS

Introduction	v
Chapter 1 Exercises	1
Chapter 2 Exercises	7
Chapter 3 Exercises	13
Chapter 4 Exercises	17
Chapter 5 Exercises	24
Chapter 6 Exercises	26
Chapter 7 Exercises	31
Chapter 8 Exercises	35
Chapter 9 Exercises	37
Chapter 10 Exercises	40
Chapter 11 Exercises	43
Sources	44
Answers to Odd-Numbered Additional Exercises	47
Section 4.4 **Control Charts**	53
Chapter 12 **Nonparametric Tests**	68

INTRODUCTION

We learn statistics by doing statistics. Working exercises is therefore perhaps the most important aspect of studying statistics. *The Basic Practice of Statistics* (BPS) offers a large number of exercises. Most follow these principles:

- Use real settings and real data.

- Ask questions that lead to a conclusion in the real setting. A number, a graph, or "reject H_0" is not a full solution to an exercise in statistics.

That is, BPS attempts to illustrate how statistical methods are used in real settings, subject of course to the constraints of a text intended for beginning students.

New exercises

This supplement to BPS provides **147 additional exercises**. Of these, 109 are traditional exercises that attempt to follow the principles just stated. Students who use software can download the new data sets, as well as data sets for BPS itself, from the BPS companion Web site at www.whfreeman.com/bps. As in BPS itself, **answers to odd-numbered exercises**, by Professor Darryl Nester of Bluffton College, are included.

Applets and applet exercises

Since the publication of BPS, W. H. Freeman has developed and made available online a set of interactive **statistical applets** that automate calculations and graphics in a way that can greatly assist learning. To use the applets, go to www.whfreeman.com/bps, click on "Statistical applets," and enter the password bpsapplets.

The applets are excellent tools for learning. No amount of written exposition, lecturing, or blackboard sketching can demonstrate (for example) the danger of influential points in regression or the behavior of confidence intervals as clearly as interactive animation. I have included here 38 **applet exercises** that guide use of the applets to gain understanding of statistical ideas. I hope that students will explore the applets on their own after completing applet exercises. Among the applets you will also find straightforward tools that for some purposes can replace calculators and statistical software. I have included notes that comment on the usefulness of applets for the relevant chapters of BPS.

An additional chapter

Finally, this supplement contains an **additional chapter** of BPS, which introduces "nonparametric" tests for use in small samples of data which clearly violate the requirements for use of common inference procedures based on normal distributions.

David S. Moore

THE BASIC PRACTICE OF STATISTICS
ADDITIONAL EXERCISES

CHAPTER 1 EXERCISES

1.1 Mutual funds. Here is a small part of a data set that describes mutual funds available to the public:

Fund	Category	Net assets ($million)	Year to date return	Largest holding
⋮				
Fidelity Low-Priced Stock	Small value	6,189	4.56%	Dallas Semiconductor
Price International Stock	International stock	9,745	−0.45%	Vodafone
Vanguard 500 Index	Large blend	89,394	3.45%	General Electric
⋮				

(a) What individuals does this data set describe?
(b) In addition to the fund's name, how many variables does the data set contain? Which of these variables are categorical and which are quantitative?
(c) What are the units of measurement for each of the quantitative variables?

1.2 House prices. The National Association of Realtors reports that the "average" sales price for existing single-family homes sold in 2000 was either $139,100 or $177,000, depending on which "average" we use.[1] Which of these numbers is the mean price and which is the median? How do you know?

1.3 A big toe problem. Hallux abducto valgus (call it HAV) is a deformation of the big toe that is not common in youth and often requires surgery. Doctors used X-rays to measure the angle (in degrees) of deformity in 38 consecutive patients under the age of 21 who came to a medical center for surgery to correct HAV. The angle is a measure of the seriousness of the deformity. Here are the data.[2] (The data set is E01-03.dat.)

```
28  32  25  34  38  26  25  18  30  26  28  13  20
21  17  16  21  23  14  32  25  21  22  20  18  26
16  30  30  20  50  25  26  28  31  38  32  21
```

Make a graph and give a numerical description of this distribution. Are there any outliers? Write a brief discussion of the shape, center, and spread of the angle of deformity among young patients needing surgery for this condition.

1.4 More on a big toe problem. The HAV angle data in the previous problem contain one high outlier. Calculate the median, the mean, and the standard deviation for the full data set and also for the 37 observations remaining when you remove the outlier. How strongly does the outlier affect each of these measures?

1.5 Returns on common stocks. The total return on a stock is the change in its market price plus any dividend payments made. Total return is usually expressed as a percent of the beginning price. Figure 1 is a histogram of the distribution of the monthly returns for all stocks listed on U.S. markets for the years 1951 to 2000 (600 months).[3] The low outlier is the market crash of October, 1987, when stocks lost more than 22% of their value in one month.

(a) Describe the overall shape of the distribution of monthly returns.
(b) What is the approximate center of this distribution? (For now, take the center to be the value with roughly half the months having lower returns and half having higher returns.)
(c) Approximately what were the smallest and largest total returns, leaving out the outlier? (This describes the spread of the distribution.)
(d) A return less than zero means that stocks lost value in that month. About what percent of all months had returns less than zero?

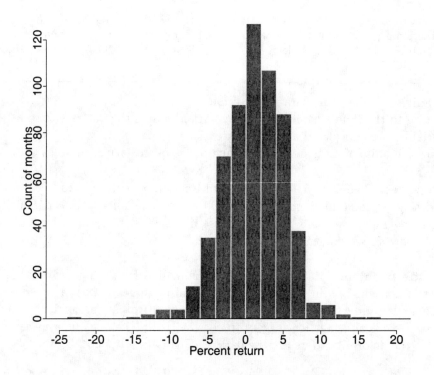

Figure 1: Monthly returns on common stocks, 1950–2000

1.6 You create the data. Create a set of 5 positive numbers (repeats allowed) that have median 10 and mean 7. What thought process did you use to create your numbers?

1.7 Where are the doctors? Table 1 gives the number of medical doctors per 100,000 people in each state. (The data set is E01-07.dat.)
(a) Why is the number of doctors per 100,000 people a better measure of the availability of health care than a simple count of the number of doctors in a state?

Table 1: Medical doctors per 100,000 population, by state (1998)

State	Doctors	State	Doctors	State	Doctors
Alabama	194	Louisiana	239	Ohio	230
Alaska	160	Maine	214	Oklahoma	166
Arizona	200	Maryland	362	Oregon	221
Arkansas	185	Massachusetts	402	Pennsylvania	282
California	244	Michigan	218	Rhode Island	324
Colorado	234	Minnesota	247	South Carolina	201
Connecticut	344	Mississippi	156	South Dakota	177
Delaware	230	Missouri	225	Tennessee	242
Florida	232	Montana	188	Texas	196
Georgia	204	Nebraska	213	Utah	197
Hawaii	252	Nevada	169	Vermont	288
Idaho	150	New Hampshire	230	Virginia	233
Illinois	253	New Jersey	287	Washington	229
Indiana	192	New Mexico	209	West Virginia	210
Iowa	171	New York	375	Wisconsin	224
Kansas	202	North Carolina	225	Wyoming	167
Kentucky	205	North Dakota	219	D.C.	702

(b) Make a graph that displays the distribution of doctors per 100,000 people. Write a brief description of the distribution. Are there any outliers? If so, can you explain them?

1.8 Where are the doctors, continued. Table 1 gives the number of medical doctors per 100,000 people in each state. Your graph of the distribution shows that the District of Columbia (D.C.) is a high outlier. Because D.C. is a city rather than a state, we will omit it here.
(a) Calculate both the five-number summary and $\bar{x}$ and s for the number of doctors per 100,000 people in the 50 states. Based on your graph, which description do you prefer?
(b) What facts about the distribution can you see in the graph that the numerical summaries don't reveal? Remember that measures of center and spread are not complete descriptions of a distribution.

1.9 How much oil? How much oil wells in a given field will ultimately produce is key information in deciding whether to drill more wells. Here are the estimated total amounts of oil recovered from 64 wells in the Devonian Richmond Dolomite area of the Michigan basin, in thousands of barrels.[4] (The data set if E01-09.dat.)

21.71	53.2	46.4	42.7	50.4	97.7	103.1	51.9
43.4	69.5	156.5	34.6	37.9	12.9	2.5	31.4
79.5	26.9	18.5	14.7	32.9	196	24.9	118.2
82.2	35.1	47.6	54.2	63.1	69.8	57.4	65.6
56.4	49.4	44.9	34.6	92.2	37.0	58.8	21.3
36.6	64.9	14.8	17.6	29.1	61.4	38.6	32.5
12.0	28.3	204.9	44.5	10.3	37.7	33.7	81.1
12.1	20.1	30.5	7.1	10.1	18.0	3.0	2.0

(a) Graph the distribution and describe its main features.
(b) Find the mean and median of the amounts recovered. Explain how the relationship between the mean and the median reflects the shape of the distribution.
(c) Give the five-number summary and explain briefly how it reflects the shape of the distribution.

1.10 NCAA rules for athletes. The National Collegiate Athletic Association (NCAA) requires Division I athletes to score at least 820 on the combined mathematics and verbal parts of the SAT exam in order to compete in their first college year. (Higher scores are required for students with poor high school grades.) In 1999, the scores of the more than one million students taking the SATs were approximately normal with mean 1017 and standard deviation 209. What percent of all students had scores less than 820?

1.11 More NCAA rules. The NCAA considers a student a "partial qualifier" eligible to practice and receive an athletic scholarship, but not to compete, if the combined SAT score is at least 720. Use the information in the previous exercise to find the percent of all SAT scores that are less than 720.

1.12 Grading managers. Many companies "grade on a bell curve" to compare the performance of their managers and professional workers. This forces the use of some low performance ratings, so that not all workers are listed as "above average." Ford Motor Company's "performance management process," for example, assigns 10% A grades, 80% B grades, and 10% C grades to the company's 18,000 managers.[5] Suppose that Ford's performance scores really are normally distributed. This year, managers with scores less than 25 received C's and those with scores above 475 received A's. What are the mean and standard deviation of the scores?

1.13 High scores on the SAT. It is possible to score higher than 800 on the SAT, but scores above 800 are reported as 800. (That is, a student can get a reported score of 800 without a perfect paper.) In 2000, the scores of men on the math part of the SAT followed a normal distribution with mean 533 and standard deviation 115. What percent of scores were above 800 (and so reported as 800)?

One-variable calculator applet exercises

The interactive applets for *The Basic Practice of Statistics* are found on the BPS companion Web site, www.whfreeman.com/bps. You can use the one-variable statistical calculator in place of a calculator or software to do both calculations ($\bar{x}$ and s and the five-number summary) and graphs (histograms and stemplots). The applet is more convenient than most calculators. It is less convenient than good software because, depending on your browser, it may be difficult to read in new data sets in one operation and to print output.

1.14 A big toe problem. Exercise 1.3 gives data on the angle of deformity for 38 young patients who require surgery to correct a deformity of their big toes. Enter these data into the calculator applet. Use the applet to do Exercise 1.3. (As a check on your date entry, there should be 38 observations with mean $\bar{x} = 25.421$. You can edit entries in the data box if you mistyped an observation.)

1.15 How histograms behave. The data set menu that accompanies the applet includes the oil well production data in Exercise 1.9. Choose these data, then click on the "Histogram" tab to see a histogram.
(a) How many classes does the applet choose to use? (You can click on the graph outside the bars to get a count of classes.)
(b) Click on the graph and drag to the left. What is the smallest number of classes you can get? What are the lower and upper bounds of each class? (Click on the bar to find out.) Make a rough sketch of this histogram.
(c) Click and drag to the right. What is the greatest number of classes you can get? How many observations does the largest class have?
(d) You see that the choice of classes changes the appearance of a histogram. Drag back and forth until you get the histogram you think best displays the distribution. How many classes did you use?

Mean and median applet exercises

1.16 Place two observations on the line by clicking below it. Why does only one arrow appear?

1.17 Place three observations on the line by clicking below it, two close together near the center of the line, and one somewhat to the right of these two.
(a) Pull the single right-most observation out to the right. (Place the cursor on the point, hold down a mouse button and drag the point.) How does the mean behave? How does the median behave? Explain briefly why each measure acts as it does.
(b) Now drag the single point to the left as far as you can. What happens to the mean? What happens to the median as you drag this point past the other two (watch carefully)?

1.18 Place 5 observations on the line by clicking below it.
(a) Add one additional observation *without changing the median*. Where is your new point?
(b) Use the applet to convince yourself that when you add yet another observation (there are now 7 in all), the median does not change no matter where you put the 7th point. Explain why this must be true.

Normal curve applet exercises

The applet allows you to do normal calculations quickly. It is somewhat limited by the number of pixels available for use, so that it can't hit every value exactly. In the exercises below, use the closest available values. In each case, *make a sketch* of the curve from the applet marked with the values you used to answer the questions asked.

1.19 The 68–95–99.7 rule for normal distributions is a useful approximation. To see how accurate the rule is, drag one flag across the other so that the applet shows the area under the curve between the two flags.
(a) Place the flags one standard deviation on either side of the mean. What is the area between these two values? What does the 68-95-99.7 rule say this area is?
(b) Repeat for locations two and three standard deviations on either side of the mean. Again compare the 68-95-99.7 rule with the area given by the applet.

1.20 How many standard deviations above and below the mean do the quartiles of any normal distribution lie? (Use the standard normal distribution to answer this question.)

1.21 In Exercise 1.12, we saw that Ford Motor Company grades its managers in such a way that the top 10% receive an A grade, the bottom 10% a C, and the middle 80% a B. Let's suppose that performance scores follow a normal distribution. How many standard deviations above and below the mean do the A/B and B/C cutoffs lie? (Use the standard normal distribution to answer this question.)

1.22 The average performance of women on the SAT, especially the math part, is lower than that of men. The reasons for this gender gap are controversial. In 2000, women's scores on the math SAT followed a normal distribution with mean 498 and standard deviation 109. The mean for men was 533. What percent of women scored higher than the male mean?

1.23 Changing the mean of a normal distribution by a moderate amount can greatly change the percent of observations in the tails. Suppose that a college is looking for applicants with SAT math scores 750 and above.
(a) In 2000, the scores of men on the math SAT followed a normal distribution with mean 533 and standard deviation 115. What percent of men scored 750 or better?
(b) Women's scores that year had a normal distribution with mean 498 and standard deviation 109. What percent of women scored 750 or better? You see that the percent of men above 750 is almost three times the percent of women with such high scores.

CHAPTER 2 EXERCISES

2.1 Age and income. How do the incomes of working-age people change with age? Because many older women have been out of the labor force for much of their lives, we look only at men between the ages of 25 and 65. Because education strongly influences income, we look only at men who have a bachelor's degree but no higher degree. A government sample survey tells us the age and income of a random sample of 5712 such men.[6] Figure 2 is a scatterplot of these data. Here is software output for regressing income on age. The line in the scatterplot is the least-squares regression line from this output.

	Coefficients	Standard Error	t Stat	P-value
Intercept	24874.3745	2637.4198	9.4313	5.75E-21
AGE	892.1135	61.7639	14.4439	1.79E-46

(a) The scatterplot in Figure 2 has a distinctive form. Why do the points fall into vertical stacks?

(b) Give some reasons why older men in this population might earn more than younger men. Give some reasons why younger men might earn more than older men. What do the data show about the relationship between age and income in the sample? Is the relationship very strong?

(c) What is the equation of the least-squares line for predicting income from age? What specifically does the slope of this line tell us?

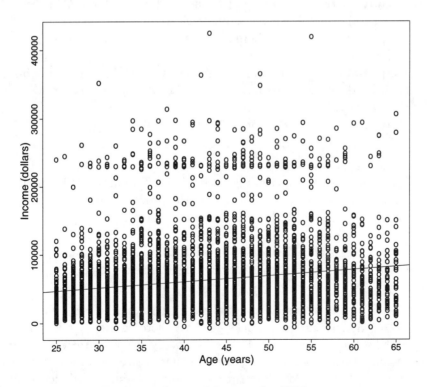

Figure 2: Age and income for 5712 men.

Table 2: Angle of deformity (degrees) for two types of foot deformity

HAV angle	MA angle	HAV angle	MA angle	HAV angle	MA angle
28	18	21	15	16	10
32	16	17	16	30	12
25	22	16	10	30	10
34	17	21	7	20	10
38	33	23	11	50	12
26	10	14	15	25	25
25	18	32	12	26	30
18	13	25	16	28	22
30	19	21	16	31	24
26	10	22	18	38	20
28	17	20	10	32	37
13	14	18	15	21	23
20	20	26	16		

2.2 Foot problems. Metatarsus adductus (call it MA) is a turning in of the front part of the foot that is common in adolescents and usually corrects itself. Hallux abducto valgus (call it HAV) is a deformation of the big toe that is not common in youth and often requires surgery. Perhaps the severity of MA can help predict the severity of HAV. Table 2 gives data on 38 consecutive patients who came to a medical center for HAV surgery.[7] (The data set is E02-02.dat.) Using X-rays, doctors measured the angle of deformity for both MA and HAV. They speculated that there is a positive correlation—more serious MA is associated with more serious HAV.
(a) Make a scatterplot of the data in Table 2. (Which is the explanatory variable?)
(b) Describe the form, direction, and strength of the relationship between MA angle and HAV angle. Are there any clear outliers in your graph?
(c) Do you think the data confirm the doctors' speculation?

2.3 Foot problems, continued.
(a) Find the equation of the least-squares regression line for predicting HAV angle from MA angle. Add this line to the scatterplot you made in the previous problem.
(b) A new patient has MA angle 25 degrees. What do you predict this patient's HAV angle to be?
(c) Does knowing MA angle allow doctors to predict HAV angle accurately? Explain your answer from the scatterplot, then calculate a numerical measure to support your finding.

2.4 Moving in step? One reason to invest abroad is that markets in different countries don't move in step. When American stocks go down, foreign stocks may go up. So an investor who holds both bears less risk. That's the theory. Now we read that "The correlation between changes in American and European share prices has risen from 0.4 in the mid-1990s to 0.8 in 2000."[8] Explain to an investor who knows no statistics why this fact reduces the protection provided by buying European stocks.

Chapter 2

2.5 Interpreting correlation. The same article that claims that the correlation between changes in stock prices in Europe and the United States was 0.8 in 2000 goes on to say that "Crudely, that means that movements on Wall Street can explain 80% of price movements in Europe." Is this true? What is the correct percent explained if $r = 0.8$?

2.6 Stocks and bonds. How is the flow of investors' money into stock mutual funds related to the flow of money into bond mutual funds? Here are data on the net new money flowing into stock and bond mutual funds in the years 1985 to 2000, in billions of dollars.[9] (The data set is E02-06.dat.) "Net" means that funds flowing out are subtracted from those flowing in. If more money leaves than arrives, the net flow will be negative. To eliminate the effect of inflation, all dollar amounts are in "real dollars" with constant buying power equal to that of a dollar in the year 2000.

Year	1985	1986	1987	1988	1989	1990	1991	1992
Stocks	12.8	34.6	28.8	−23.3	8.3	17.1	50.6	97.0
Bonds	100.8	161.8	10.6	−5.8	−1.4	9.2	74.6	87.1

Year	1993	1994	1995	1996	1997	1998	1999	2000
Stocks	151.3	133.6	140.1	238.2	243.5	165.9	194.3	309.0
Bonds	84.6	−72.0	−6.8	3.3	30.0	79.2	−6.2	−48.0

(a) Make a scatterplot with cash flow into stock funds as the explanatory variable.
(b) Find the least-squares line for predicting net bond investments from net stock investments. Add this line to your plot.
(c) What do the data suggest?

2.7 Illiteracy. Literacy is an important contributor to the economic and social development of a country. The data set E02-07.dat gives the percent of men and women at least 15 years old who were illiterate in 138 developing nations in the year 2000.[10] Use software to analyze these data.
(a) Make a scatterplot of female illiteracy versus male illiteracy. Because schooling for women often lags behind schooling for men, we take the percent of illiterate males as our explanatory variable. Describe the form, direction, and strength of the relationship, giving a numerical measure of the strength.
(b) Find the equation of the least-squares regression line for predicting female illiteracy from male illiteracy and draw this line on your plot. Which countries have the largest positive residual (female illiteracy is higher than predicted) and negative residual (female illiteracy is lower than predicted)? Which countries have the highest rates of illiteracy for both men and women?

2.8 Mutual fund performance. Many mutual funds compare their performance with that of a benchmark, an index of the returns on all securities of the kind the fund buys. The Vanguard International Growth Fund, for example, takes as its benchmark the Morgan Stanley EAFE (Europe, Australasia, Far East) index of overseas stock market performance. Here are the percent returns for the fund and for the EAFE from 1982 (the first full year of the fund's existence) to 2000.[11] (The data set is E02-08.dat.)

Year	Fund	EAFE	Year	Fund	EAFE
1982	5.27	−0.86	1992	−5.79	−11.85
1983	43.08	24.61	1993	44.74	32.94
1984	−1.02	7.86	1994	0.76	8.06
1985	56.94	56.72	1995	14.89	11.55
1986	56.71	69.94	1996	14.65	6.36
1987	12.48	24.93	1997	4.12	2.06
1988	11.61	28.59	1998	16.93	20.33
1989	24.76	10.80	1999	26.34	27.30
1990	−12.05	−23.20	2000	−8.60	−13.96
1991	4.74	12.50			

(a) Make a scatterplot suitable for predicting fund returns from EAFE returns. Is there a clear straight-line pattern? How strong is this pattern? (Give a numerical measure.) Are there any extreme outliers from the straight-line pattern?

(b) Find the equation of the least-squares regression line for predicting fund return from EAFE return. In a year when overseas markets as a group return 10% (as measured by the EAFE), what do you predict to be the return for the fund?

Two-variable calculator applet exercises

The interactive applets for *The Basic Practice of Statistics* are found on the BPS companion Web site, www.whfreeman.com/bps. You can use the two-variable statistical calculator in place of a calculator or software to do both calculations (means and standard deviations, correlation, least-squares line) and graphs (scatterplot and residual plot). The applet is more convenient than most calculators. It is less convenient than good software because, depending on your browser, it may be difficult to read in new data sets in one operation and to print output.

2.9 Illiteracy. The data set menu that accompanies the two-variable calculator includes the male and female illiteracy rates for 138 countries, described in Exercise 2.7. Choose this data set and use the calculator to do Exercise 2.7.

2.10 Mutual fund performance. Exercise 2.8 gives data on the percent return for the Vanguard International Growth Fund and its benchmark index of overseas stock market performance, the Morgan Stanley EAFE, for 19 years. Enter these data into the calculator applet and use the applet to do Exercise 2.8. (As a check on your data entry, the correlation should be $r = 0.8983$. You can edit entries in the data box.)

Correlation and regression applet exercises

This interactive applet shows how the correlation r and the least-squares regression line respond to changes in a scatterplot. No amount of talking or writing can show so clearly how these statistical measures behave. The following exercises point to some important

Chapter 2   11

facts, but I hope that you will experiment a bit to gain some feeling for correlation and regression. To erase an entire scatterplot and start over, click on the trash can.

2.11 Match the correlation. You are going to make scatterplots with 10 points that have correlation close to 0.7. The lesson is that many patterns can have the same correlation. Always plot your data before you trust a correlation.
(a) Stop after adding the first two points. What is the value of the correlation? Why does it have this value?
(b) Make a lower left to upper right pattern of 10 points with correlation about $r = 0.7$. (You can drag points up or down to adjust r after you have 10 points.) Make a rough sketch of your scatterplot.
(c) Make another scatterplot with 9 points in a vertical stack at the left of the plot. Add one point far to the right and move it until the correlation is close to 0.7. Make a rough sketch of your scatterplot.
(d) Make yet another scatterplot with 10 points in a curved pattern that starts at the lower left, rises to the right, then falls again at the far right. Adjust the points up or down until you have a quite smooth curve with correlation close to 0.7. Make a rough sketch of this scatterplot also.

2.12 Is regression useful? In the previous exercise, you created three scatterplots having correlation about $r = 0.7$ between the horizontal variable x and the vertical variable y. Correlation $r = 0.7$ is considered reasonably strong in many areas of scientific work. Because there is a reasonably strong correlation, we might use a regression line to predict y from x. In which of your three scatterplots does it make sense to use a straight line for prediction?

2.13 Influence on correlation. Click on the scatterplot to create a group of 10 points in the lower left corner of the scatterplot with a strong straight-line pattern (correlation about 0.9).
(a) Add one point at the upper right that is in line with the first 10. How does the correlation change?
(b) Drag this last point down until it is opposite the group of 10 points. How small can you make the correlation? Can you make the correlation negative? You see that a single outlier can greatly strengthen or weaken a correlation. Always plot your data to check for outlying points.

2.14 Influence in regression. As in the previous exercise, create a group of 10 points in the lower left corner of the scatterplot with a strong straight-line pattern (correlation at least 0.9). Click the "Show least-squares line" box to display the regression line.
(a) Add one point at the upper right that is far from the other 10 points but exactly on the regression line. Why does this outlier have no effect on the line even though it changes the correlation?
(b) Now drag this last point down until it is opposite the group of 10 points. You see that one end of the least-squares line chases this single point, while the other end remains near the middle of the original group of 10. What about the last point makes it so influential?

2.15 Guessing a regression line. Click on the scatterplot to create a group of 15 to 20 points from lower left to upper right with a clear positive straight-line pattern (correlation around 0.7). Click the "Draw line" button and use the mouse (right-click and drag) to draw a line through the middle of the cloud of points from lower left to upper right. Note the "thermometer" above the plot. The red portion is the sum of the squared vertical distances from the points in the plot to the least-squares line. The green portion is the "extra" sum of squares for your line—it shows by how much your line misses the smallest possible sum of squares.

(a) You drew a line by eye through the middle of the pattern. Yet the right-hand part of the bar is probably almost entirely green. What does that tell you?

(b) Now click the "Show least-squares line" box. Is the slope of the least-squares line smaller (the new line is less steep) or larger (line is steeper) than that of your line? If you repeat this exercise several times, you will consistently get the same result. The least-squares line minimizes the *vertical* distances of the points from the line. It is *not* the line through the "middle" of the cloud of points. This is one reason why it is hard to draw a good regression line by eye.

CHAPTER 3 EXERCISES

3.1 Instant opinion. The Harris/Excite instant poll can be found online at `news.excite.com/news/poll`. The question appears on the screen, and you simply click buttons to vote "Yes," "No," or "Don't Know." On January 25, 2000, the question was "Should female athletes be paid the same as men for the work they do?" In all, 13,147 (44%) said "Yes," another 15,182 (50%) said "No," and the remaining 1448 said "Don't know."
(a) What is the sample size for this poll?
(b) That's a much larger sample than standard sample surveys. In spite of this, we can't trust the result to give good information about any clearly defined population. Why?
(c) It is still true that more men than women use the Web. How might this fact affect the poll results?

3.2 Dealing with pain. Health care providers are giving more attention to relieving the pain of cancer patients. An article in the journal *Cancer* surveyed a number of studies and concluded that controlled-release morphine tablets, which release the pain killer gradually over time, are more effective that giving standard morphine when the patient needs it.[12] The "methods" section of the article begins: "Only those published studies that were controlled (i.e., randomized, double blind, and comparative), repeated-dose studies with CR morphine tablets in cancer pain patients were considered for this review." Explain the terms in parentheses to someone who know nothing about medical trails.

3.3 Protecting ultramarathon runners. An ultramarathon, as you might guess, is a foot race longer than the 26.2 miles of a marathon. Runners commonly develop respiratory infections after an ultramarathon. Will taking 600 milligrams of vitamin C daily reduce these infections? Researchers randomly assigned ultramarathon runners to receive either vitamin C or a placebo. Separately, they also randomly assigned these treatments in a group of non-runners the same age as the runners. All subjects were watched for 14 days after the big race to see if infections developed.[13]
(a) What is the name for this experimental design?
(b) Use a diagram to outline the design.

3.4 Exercising to lose weight. We all know that regular exercise (combined with a sensible diet) is a key to shedding those extra pounds. Experience shows that overweight people find it tough to keep exercising. Perhaps they will do better with several short sessions each day rather than one longer session. Perhaps having exercise equipment at home will help. An experiment looked at these issues. The subjects were women aged 25 to 45 whose weights were 20% to 75% higher than ideal. The study report says:[14]

> Subjects were randomly assigned to 1 of 3 groups. All groups were prescribed a similar volume of exercise. The 3 groups differed in the way the exercise was prescribed. ...
>
> **Long-Bout Exercise Group** *Forty-nine subjects were instructed to exercise 5 d/wk; duration progressed from 20 min/d ... to 40 min/d ... Participants performed the exercise in one long bout.*

Short-Bout Exercise Group *Fifty-one subjects were instructed to exercise 5 d/wk ...However, rather than exercising continuously for the prescribed duration, subjects were instructed to divide the exercise into multiple 10-minute bouts that were performed at convenient times throughout the day.*

Short-Bout Plus Exercise Equipment Group *The exercise prescription was identical to the exercise prescribed for the short-bout group ...The 48 subjects in this group were also provided with motorized home treadmills.*

The researchers recorded weight, fitness, and whether the subject continued the exercise program.
(a) Use a diagram to outline the design of this experiment.
(b) How many subjects are there in all? Use Table A starting at line 114 to choose the first 10 subjects for the long-bout group.

3.5 A telephone survey. The 1998 National Gun Policy Survey, carried out by the University of Chicago's National Opinion Research Center, asked respondents' opinions about government regulation of firearms. A report from the survey says, "Participating households were identified through random-digit dialing; the respondent in each household was selected by the most-recent-birthday method."[15]
(a) What is "random-digit dialing?" Why is it a practical method for obtaining (almost) an SRS of households?
(b) The survey wants the opinion of an individual adult. Several adults may live in a household. In that case, the survey interviewed the adult with the most recent birthday. Why is this preferable to simply interviewing the person who answers the phone?

3.6 Are you sure? Late in 1996, Spain's Centro de Investigaciones Sociológicos carried out a sample survey on the attitudes of Spaniards toward private business and state intervention in the economy.[16] Of the 2496 adults interviewed, 72% agreed that, "Employees with higher performance must get higher pay." On the other hand, 71% agreed that, "Everything a society produces should be distributed among its members as equally as possible and there should be no major differences." Use these conflicting results as an example in a short explanation of why opinion polls often fail to reveal public attitudes clearly.

3.7 Reducing risky sex. The National Institutes of Mental Health (NIMH) wants to know whether intense education about the risks of AIDS will help change the behavior of people who now report sexual activities that put them at risk of infection. NIMH investigators screened 38,893 people to identify 3706 suitable subjects. The subjects were assigned to a control group (1855 people) or an intervention group (1851 people). The control group attended a one-hour AIDS education session; the intervention group attended seven single-sex discussion sessions, each lasting 90 to 120 minutes. After 12 months, 64% of the intervention group and 52% of the control group said they used condoms. (None of the subjects used condoms regularly before the study began.)[17]
(a) Because none of the subjects used condoms when the study started, we might just offer the intervention sessions and find that 64% used condoms 12 months after the sessions. Explain why this greatly overstates the effectiveness of the intervention.
(b) Outline the design of this experiment.

Chapter 3

(c) You must randomly assign 3706 subjects. How would you label them? Use line 119 of Table B to choose the first 5 subjects for the intervention group.

3.8 Growing trees faster. The concentration of carbon dioxide (CO_2) in the atmosphere is increasing rapidly due to our use of fossil fuels. Because plants use CO_2 to fuel photosynthesis, more CO_2 may cause trees and other plants to grow faster. An elaborate apparatus allows researchers to pipe extra CO_2 to a 30-meter circle of forest. We want to compare the growth in base area of trees in treated and untreated areas to see if extra CO_2 does in fact increase growth. We can afford to treat three circular areas.[18]
(a) Describe the design of a completely randomized experiment using 6 well-separated 30-meter circular areas in a pine forest. Sketch the circles and carry out the randomization your design calls for.
(b) Areas within the forest may differ in soil fertility. Describe a matched pairs design using three pairs of circles that will reduce the extra variation due to different fertility. Sketch the circles and carry out the randomization your design calls for.

3.9 Keeping warm during surgery (EESEE). Surgery patients are often cold because the operating room is kept cool and the body's temperature regulation is disturbed by anesthetics. Will warming patients to maintain normal body temperature reduce infections after surgery? In one experiment, patients undergoing colon/rectal surgery received intravenous fluids from a warming machine and were covered with a blanket through which air circulated. In some patients, the fluid and the air were warmed; in others, they were not. The patients received identical treatment in all other respects.[19]
(a) To simplify the setup of the study, we might warm the fluids and air blanket for one operating team and not for another doing the same kind of surgery. Why might this design result in bias?
(b) Outline the design of a randomized comparative experiment for this study.
(c) The operating team did not know whether fluids and air blanket were heated, nor did the doctors who followed the patients after surgery. What is this practice called? Why was it used here?

Simple random sample applet exercises

The interactive applets for *The Basic Practice of Statistics* are found on the BPS companion Web site, www.whfreeman.com/bps. The simple random sample applet can choose an SRS of any size up to $n = 40$ from a population of any size up to 500.

3.10 Sampling retail outlets. Exercise 3.9 on page 174 of BPS asks you to choose an SRS of 10 from the 440 retail outlets in New York that sell your product. Use the applet to choose this sample. Which outlets were chosen? (That was faster than using Table B.)

3.11 Testing a breakfast food. Because experimental randomization chooses SRSs of the subjects, we can use the applet here as well as for sampling problems. Example 3.12 on page 190 of BPS describes a randomized comparative experiment in which 30 rats are assigned at random to a treatment group of 15 and a control group of 15. Use the applet to choose the 15 rats for the treatment group. Which rats did you choose? The remaining 15 rats make up the control group.

3.12 Conserving energy. The applet allows you to randomly assign subjects to more than two groups without difficulty. Example 3.13 on page 191 of BPS describes a randomized comparative experiment in which 60 houses are randomly assigned to three groups of 20.
(a) Use the applet to choose an SRS of 20 out of 60 houses to form the first group. Which houses are in this group?
(b) The "Population hopper" now contains the 40 houses that were not chosen, in scrambled order. Click "Sample" again to choose an SRS of 20 of these remaining houses to make up the second group. Which houses were chosen?
(c) The 20 houses remaining in the "Population hopper" form the third group. Which houses are these?

3.13 Randomization avoids bias. Suppose that the 15 even-numbered rats among the 30 rats available in the setting of Exercise 3.11 are (unknown to the experimenters) a fast-growing variety. We hope that these rats will be roughly equally distributed between the two groups. Take 10 samples of size 15 from the 30 rats. (Be sure to click "Reset" after each sample.) Record the counts of even-numbered rats in each of your 10 samples. You see that there is considerable chance variation, but no systematic bias in favor of one or the other group in assigning the fast-growing rats. Larger samples from larger population will on the average do a better job of making the two groups equivalent.

CHAPTER 4 EXERCISES

4.1 Measuring unemployment. The Bureau of Labor Statistics announces that last month it interviewed all members of the labor force in a sample of 50,000 households; **4.5%** of the people interviewed were unemployed. Is this number a parameter or a statistic? Why?

4.2 Republican voters. Voter registration records show that **68%** of all voters in Indianapolis are registered as Republicans. To test a random digit dialing device, you use the device to call 150 randomly chosen residential telephones in Indianapolis. Of the registered voters contacted, **73%** are registered Republicans. Is each of the boldface numbers a parameter or a statistic? Why?

4.3 Preparing for the GMAT. A company that offers courses to prepare would-be MBA students for the GMAT examination has the following information about its customers: 20% are currently undergraduate students in business; 15% are undergraduate students in other fields of study; 60% are college graduates who are currently employed; and 5% are college graduates who are not employed.
(a) Is this a legitimate assignment of probabilities to customer backgrounds? Why?
(b) What percent of customers are currently undergraduates?

4.4 Race and ethnicity. The 2000 census allowed each person to choose one or more from of a long list of races. That is, in the eyes of the Census Bureau, you belong to whatever race or races you say you belong to. "Hispanic/Latino" is a separate category; Hispanics may be of any race. If we choose a resident of the United States at random, the 2000 census gives these probabilities:

	Hispanic	Not Hispanic
Asian	0.000	0.036
Black	0.003	0.121
White	0.060	0.691
Other	0.062	0.027

(a) Verify that this is a legitimate assignment of probabilities.
(b) What is the probability that a randomly chosen American is Hispanic?
(c) Non-Hispanic whites are the historical majority in the United states. What is the probability that a randomly chosen American is not a member of this group?

4.5 Tetrahedral dice. Psychologists sometimes use tetrahedral dice to study our intuition about chance behavior. A tetrahedron is a pyramid (think of Egypt) with four identical faces, each a triangle with all sides equal in length. Label the four faces of a tetrahedral die with 1, 2, 3, and 4 spots. Give a probability model for rolling such a die and recording the number of spots on the down face. Explain why you think your model is at least close to correct.

4.6 Playing "pick four." The "pick four" games in many state lotteries announce a four-digit winning number each day. The winning number is essentially a four-digit group

from a table of random digits. You win if your choice matches the winning digits. Suppose your chosen number is 5974.
(a) What is the probability that your number matches the winning number exactly?
(b) What is the probability that your number matches the digits in the winning number *in any order*?

4.7 More tetrahedral dice. Tetrahedral dice are described in Exercise 4.5. Give a probability model for rolling two such dice. That is, write down all possible outcomes and give a probability to each. (Example 4.4 and Figure 4.2 in BPS may help you.) What is the probability that the sum of the down faces is 5?

4.8 Playing "pick four," continued. The Wisconsin version of "pick four" pays out $5000 on a $1 bet if your number matches the winning number exactly. It pays $200 on a $1 bet if the digits in your number match those of the winning number in any order. You choose which of these two bets to make. On the average over many bets, your winnings will be

$$\text{mean amount won} = \text{payout amount} \times \text{probability of winning}$$

What is this mean payout for these two bets? Is one of the two bets a better choice?

4.9 An edge in "pick four." Exercise 4.6 describes "pick four" lottery games. Some states (New Jersey, for example) use the "pari-mutuel system" in which the total winnings are divided among all players who matched the winning digits. That suggests a way to get an edge. Suppose you choose to try to match the winning number exactly.
(a) The winning number might be, for example, either 2873 or 8888. Explain why these two outcomes have exactly the same probability.
(b) It is likely that fewer people will choose one of these numbers than the other, because it "doesn't look random." You prefer the less popular number because you will win more if fewer people share a winning number. Which of these two numbers do you prefer?

4.10 Polling women. Suppose that 47% of all adult women think they do not get enough time for themselves. An opinion poll interviews 1025 randomly chosen women and records the sample proportion who don't feel they get enough time for themselves. This statistic will vary from sample to sample if the poll is repeated. The sampling distribution is approximately normal with mean 0.47 and standard deviation about 0.016. Sketch this normal curve and use it to answer the following questions.
(a) The truth about the population is 0.47. In what range will the middle 95% of all sample results fall?
(b) What is the probability that the poll gets a sample in which fewer than 45% say they do not get enough time for themselves?

4.11 Will you have an accident? The probability that a randomly chosen driver will be involved in an accident in the next year is about 0.2. This is based on the proportion of millions of drivers who have accidents. "Accident" includes things like crumpling a fender in your own driveway, not just highway accidents.
(a) What do you think is your own probability of being in an accident in the next year? This is a *personal probability* that rests on your own assessment of chance, not on many

Chapter 4   19

repeated trials.
(b) Give some reasons why your personal probability might be a more accurate prediction of your "true chance" of having an accident than the probability for a random driver.
(c) Almost everyone says their personal probability is lower than the random driver probability. Why do you think this is true?

4.12 What probability doesn't say. The probability of a head in tossing a coin is 1/2. This means that as we make more tosses, the *proportion* of heads will eventually get close to 0.5. It does not mean that the *count* of heads will get close to 1/2 the number of tosses. To see why, imagine that the proportion of heads is 0.51 in 100 tosses, 1000 tosses, 10,000 tosses, and 100,000 tosses of a coin. How many heads came up in each set of tosses? How close is the number of heads to half the number of tosses?

4.13 A sampling distribution. We want to know what percent of American adults approve of legal gambling. This population proportion p is a parameter. To estimate p, take an SRS and find the proportion $\hat{p}$ in the sample who approve of gambling. If we take many SRSs of the same size, the proportion $\hat{p}$ will vary from sample to sample. The distribution of its values in all SRSs is the sampling distribution of this statistic.

Figure 3 on the following page is a small population. Each circle represents an adult. The circles containing dots are people who disapprove of legal gambling, and the empty circles are people who approve. You can check that 60 of the 100 circles are empty, so in this population the proportion who approve of gambling is $p = 60/100 = 0.6$.
(a) Label the population 00 to 99 left-to-right across the rows, starting at the top left. Use line 101 of Table B to draw an SRS of size 5. What is the proportion $\hat{p}$ of the people in your sample who approve of gambling?
(b) Take 9 more SRSs of size 5 (10 in all), using lines 102 to 110 of Table B, a different line for each sample. You now have 10 values of the sample proportion $\hat{p}$. What are they?
(c) Because your samples have only 5 people, the only values $\hat{p}$ can take are 0/5, 1/5, 2/5, 3/5, 4/5, and 5/5. That is, $\hat{p}$ is always 0, 0.2, 0.4, 0.6, 0.8, or 1. Mark these numbers on a line and make a histogram of your 10 results by putting a bar above each number to show how many samples had that outcome. (You have begun to construct the sampling distribution of $\hat{p}$, though of course 10 samples is a small start.)
(d) Taking samples of size 5 from a population of size 100 is not a practical setting, but let's look at your results anyway. How many of your 10 samples estimated the population proportion $p = 0.6$ exactly correctly? Is the true value 0.6 roughly in the center of your sample values? Explain why 0.6 would be in the center of the sample values if you took a large number of samples.

4.14 Insurance. The idea of insurance is that we all face risks that are unlikely but carry high cost. Think of a fire destroying your home. So we form a group to share the risk: We all pay a small amount, and the insurance policy pays a large amount to those few of us whose homes burn down. An insurance company looks at the records for millions of homeowners and sees that the mean loss from fire in a year is $\mu = \$250$ per person. (Most of us have no loss, but a few lose their homes. The $250 is the average loss.) The company plans to sell fire insurance for $250 plus enough to cover its costs and profit. Explain clearly why it would be stupid to sell only 12 policies. Then explain why selling thousands of such policies is a safe business.

4.15 More on insurance. In fact, the insurance company sees that in the entire population of homeowners, the mean loss from fire is $\mu = \$250$ and the standard deviation of the loss is $\sigma = \$300$. The distribution of losses is strongly right-skewed: Many policies have $0 loss, but a few have large losses. If the company sells 10,000 policies, what is the approximate probability that the average loss will be greater than $260?

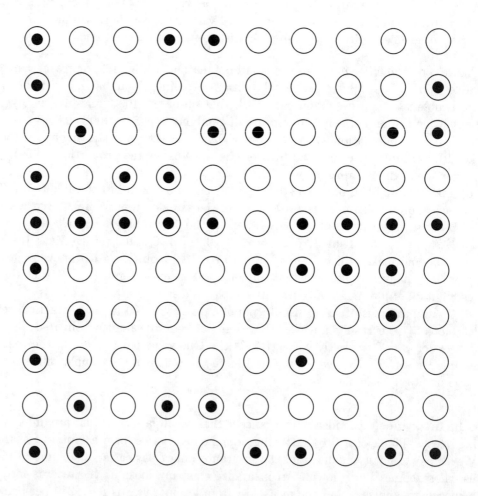

Figure 3: A population in which 60% approve of legal gambling.

Probability applet exercises

The interactive applets for *The Basic Practice of Statistics* are found on the BPS companion Web site, www.whfreeman.com/bps. The coin-tossing in the probability applet is a model for any setting with repeated independent success-or-failure trials, each of which has the same probability of a success. You can simulate up to 40 trials at once.

4.16 The nature of probability. Suppose that you toss a balanced coin very many times. To simulate this, set the "Probability of a head" in the applet to 0.5. The applet allows no more than 40 tosses at once, but you can add 40 more by clicking "Toss" again. Check the "Show true probability" box to display the probability 0.5 on the graph.
(a) Simulate 200 tosses of a coin by clicking "Toss" five times. The graph shows how the proportion of heads changes as you make more tosses. What was this proportion after 200 tosses? Make a rough sketch of the graph that displays how the proportion eventually gets close to the probability 0.5.
(b) Click "Reset" and do another 200 tosses. What was the proportion of heads in these 200 tosses? Sketch the graph again. The two graphs, representing two sets of 200 tosses, often look very different. What they have in common is that the proportion of heads eventually gets close to the probability 0.5.

4.17 What probability doesn't say. The idea of probability is that the *proportion* of heads in many tosses of a balanced coin eventually gets close to 0.5. But does the actual *count* of heads get close to one-half the number of tosses? Let's find out. Set the "Probability of heads" in the applet to 0.5 and the number of tosses to 40. You can extend the number of tosses by clicking "Toss" again to get 40 more. Don't click "Reset" during this exercise.
(a) After 40 tosses, what is the proportion of heads? What is the count of heads? What is the difference between the count of heads and 20 (one-half the number of tosses)?
(b) Keep going to 120 tosses. Again record the proportion and count of heads and the difference between the count and 60 (half the number of tosses).
(c) Keep going. Stop at 240 tosses and again at 480 tosses to record the same facts. Although it may take a long time, the laws of probability say that the proportion of heads will always get close to 0.5 and also that the difference between the count of heads and half the number of tosses will always grow without limit.

4.18 Not a great bet. In Exercise 4.6 you found the probability that the winning number in a "pick-four" lottery matches the digits in your number in any order. Enter this probability of winning as the "Probability of heads" in the applet. Enter 31 as the number of tosses. This represents a bet every day for a month. Simulate a month's play. Keep playing every day, a month at a time (click "Reset" to start a new month) until you win. You will often wait a long time to win!

4.19 A sampling distribution. You can use the probability applet to speed up and improve Exercise 4.13. You have a population in which 60% of the individuals approve of legal gambling. You want to take many small samples from this population to observe how the sample proportion who approve of gambling varies from sample to sample. Set the "Probability of heads" in the applet to 0.6 and the number of tosses to 5. This simulates an SRS of size 5 from a very large population, not just 100 individuals as in Exercise 4.13. By alternating between "Toss" and "Reset" you can take many samples quickly. (a) Take 50 samples, recording the number of heads (that is, the number in the sample who approve of gambling) in each sample. Make a histogram of the 50 sample proportions.

(b) Another population contains only 20% who approve of legal gambling. Take 50 samples of size 5 from this population, record the number in each sample who approve, and make a histogram of the 50 sample proportions. How do the centers of your two histograms reflect the differing truths about the two populations?

Expected value applet exercise

4.20 The law of large numbers. Suppose that you roll two balanced dice and look at the spots on the up faces. There are 36 possible outcomes, displayed in Figure 4.2 on page 221 of BPS. Because the dice are balanced, all 36 outcomes are equally likely. Add the spots on the up faces. The average of the 36 totals is 7. This is the population mean μ for the idealized population that contains the results of rolling two dice forever. (The mean is also called the "expected value," which explains the name of the applet. We do not expect to get the value μ on one roll, so the term is a bit misleading.) The law of large numbers says that the average $\bar{x}$ from a finite number of rolls gets closer and closer to 7 as we do more and more rolls.

(a) Click "More dice" in the expected value applet once to get two dice. Click "Show mean" to see the mean 7 on the graph. Leaving the number of rolls at 1, click "Roll dice" three times. Note the count of spots for each roll (what were they?) and the average for the three rolls. You see that the graph displays at each point the average number of spots for all rolls up to the last one. Now you understand the display.

(b) Set the number of rolls to 100 and click "Roll dice." The applet rolls the two dice 100 times. The graph shows how the average count of spots changes as we make more rolls. That is, the graph shows $\bar{x}$ as we continue to roll the dice. Make a rough sketch of the final graph.

(c) Repeat your work from (b). Click "Reset" to start over, then roll two dice 100 times. Make a sketch of the final graph of the mean $\bar{x}$ against the number of rolls. Your two graphs will often look very different. What they have in common is that the average eventually gets close to the population mean $\mu = 7$. The law of large numbers says that this will *always* happen if you keep on rolling the dice.

4.21 What's the mean? Suppose that you roll three balanced dice. We wonder what the mean number of spots on the up faces of the three dice is. The law of large numbers says that we can find out by experience: Roll three dice many times, and the actual average number of spots will eventually approach the mean. Set up the applet to roll three dice. Don't click "Show mean" yet. Roll the dice until you are confident you know the mean quite closely, then click "Show mean" to verify your discovery. What is the mean? Make a rough sketch of the path the averages $\bar{x}$ followed as you kept adding more rolls.

Simple random sample applet exercise

4.22 A sampling distribution. We can use the simple random sample applet to help grasp the idea of a sampling distribution. Form a population labeled 1 to 100. We will choose an SRS of 10 of these numbers. That is, in this exercise, the numbers themselves are the population, not just labels for 100 individuals. The mean of the whole numbers 1 to 100 is $\mu = 50.5$. This is the population mean.

(a) Use the applet to choose an SRS of size 10. Which 10 numbers were chosen? What is their mean? This is the sample mean $\bar{x}$.

(b) Although the population and its mean $\mu = 50.5$ remain fixed, the sample mean changes as we take more samples. Take another SRS of size 10. (Use the "Reset" button to return to the original population before taking the second sample.) What are the 10 numbers in your sample? What is their mean? This is another value of $\bar{x}$.

(c) Take 8 more SRSs from this same population and record their means. You now have 10 values of the sample mean $\bar{x}$ from 10 SRSs of the same size from the same population. Make a histogram of the 10 values and mark the population mean $\mu = 50.5$ on the horizontal axis. Are your 10 sample values roughly centered at the population value μ? (If you kept going forever, your $\bar{x}$-values would form the sampling distribution of the sample mean; the population mean μ would indeed be the center of this distribution.)

CHAPTER 5 EXERCISES

5.1 Race and ethnicity. The 2000 census allowed each person to choose one or more from of a long list of races. That is, in the eyes of the Census Bureau, you belong to whatever race or races you say you belong to. "Hispanic/Latino" is a separate category; Hispanics may be of any race. If we choose a resident of the United States at random, the 2000 census gives these probabilities:

	Hispanic	Not Hispanic
Asian	0.000	0.036
Black	0.003	0.121
White	0.060	0.691
Other	0.062	0.027

(a) What is the probability that a randomly chosen person is white?
(b) You know that the person chosen is Hispanic. What is the conditional probability that this person is white?

5.2 More on race and ethnicity.
(a) What is the probability that a randomly chosen American is Hispanic?
(b) You know that the person chosen is black. What is the conditional probability that this person is Hispanic?

5.3 At the gym. Many conditional probability calculations are just common sense made automatic. For example, 10% of adults belong to health clubs, and 40% of these health club members go to the club at least twice a week. What percent of all adults go to a health club at least twice a week? Write the information given in terms of probabilities and use the general multiplication rule.

5.4 A hot stock. You purchase a hot stock for $1000. The stock either gains 30% or loses 25% each day, and its behaviors on consecutive days are independent of each other. You plan to sell the stock after two days. What are the possible values of the stock after two days, and what is the probability for each value? (*Hint*: Remember that the value is multiplied by 1.30 on a day of 30% increase and multiplied by 0.75 on a day of 25% loss.)

5.5 Should you invest? Consider the hot stock of Exercise 5.4.
(a) What is the probability that the stock is worth more after two days than the $1000 you paid for it? You see that you will usually lose money if you pay $1000 for this stock.
(b) The *mean value* of the stock after two days turns out to be about $1050. The law of large numbers says that on the average over many such $1000 investments you will come out about $50 ahead. Make a probability histogram of the distribution of possible values from Exercise 5.4. Use what you know about the behavior of means to explain briefly why the mean is much larger than most of the possible values.

5.6 Teen-age drivers. An insurance company has the following information about drivers aged 16 to 18 years: 20% are involved in accidents each year; 10% in this age group are A students; among those involved in an accident, 5% are A students.

Chapter 5

(a) Let A be the event that a young driver is an A student and C the event that a young driver is involved in an accident this year. State the information given in terms of probabilities and conditional probabilities for the events A and C.
(b) What is the probability that a randomly chosen young driver is an A student and is involved in an accident?

5.7 More on teen-age drivers. Use your work from Exercise 5.6 to find the percent of A students who are involved in accidents. (Start by expressing this as a conditional probability.)

5.8 Preparing for the GMAT. A company that offers courses to prepare would-be MBA students for the GMAT examination finds that 40% of its customers are currently undergraduate students and 60% are college graduates. After completing the course, 50% of the undergraduates and 70% of the graduates achieve scores of at least 600 on the GMAT.
(a) What percent of customers are undergraduates *and* score at least 600? What percent of customers are graduates *and* score at least 600?
(b) What percent of all customers score at least 600 on the GMAT?

5.9 Screening job applicants. A company retains a psychologist to assess whether job applicants are suited for assembly-line work. The psychologist classifies applicants as A (well suited), B (marginal), or C (not suited). The company is concerned about event D: an employee leaves the company within a year of being hired. Data on all people hired in the past five years gives these probabilities:

$$P(A) = 0.4 \qquad P(B) = 0.3 \qquad P(C) = 0.3$$
$$P(A \text{ and } D) = 0.1 \qquad P(B \text{ and } D) = 0.1 \qquad P(C \text{ and } D) = 0.2$$

Sketch a Venn diagram of the events A, B, C, and D and mark on your diagram the probabilities of all combinations of psychological assessment and leaving (or not) within a year. What is $P(D)$, the probability that an employee leaves within a year?

Probability applet exercise

The interactive applets for *The Basic Practice of Statistics* are found on the BPS companion Web site, www.whfreeman.com/bps. The coin-tossing in the probability applet simulates independent success/failure outcomes. The random count of heads (or tails) in the specified number of tosses therefore follows a binomial distribution.

5.10 Inspecting switches. Example 5.10 on page 273 of BPS concerns the count of bad switches in inspection samples of size 10. The count has the binomial distribution with $n = 10$ and $p = 0.1$. Set these values for the number of tosses and probability of heads in the probability applet. The example calculates that the probability of getting a sample with exactly 1 bad switch is 0.3874. Of course, when we inspect only a few lots, the proportion of samples with exactly 1 bad switch will differ from this probability. Click "Toss" and "Reset" repeatedly to simulate inspecting 20 lots. Record the number of bad switches (the count of heads) in each of the 20 samples. What proportion of the 20 lots had exactly 1 bad switch? Remember that probability tells us only what happens in the long run.

CHAPTER 6 EXERCISES

6.1 Protecting ultramarathon runners. Exercise 3.3 describes an experiment designed to learn whether taking vitamin C reduces the incidence of respiratory infections among ultramarathon runners. The report of the study said:

> Sixty-eight percent of the runners in the placebo group reported the development of symptoms of upper respiratory tract infection after the race; this was significantly more ($P < 0.01$) than that reported by the vitamin C-supplemented group (33%).

(a) Explain to someone who knows no statistics why "significantly more" means there is good reason to think that vitamin C works.
(b) Now explain more exactly: What does $P < 0.01$ mean?

6.2 Protecting ultramarathon runners, continued. It is possible for a result to be statistically significant, but so small that it is of no practical interest. How can this happen? Do you think this happens in the study of the previous exercise?

6.3 Cell phones and brain cancer. Could the radiation from cell phones be harmful to users? Many studies have found little or no connection between using cell phones and various illnesses. Here is part of a news account of one study:[20]

> A hospital study that compared brain cancer patients and a similar group without brain cancer found no statistically significant association between cell phone use and a group of brain cancers known as gliomas. But when 20 types of glioma were considered separately an association was found between phone use and one rare form. Puzzlingly, however, this risk appeared to decrease rather than increase with greater mobile phone use.

This is a pretty weak study. Let's dissect it.
(a) Explain why this study is not an experiment.
(b) What does "no statistically significant association" mean in plain language?
(c) Why are you not surprised that in 20 separate tests for 20 types of cancer, one was significant at the 5% level?

6.4 Ages of presidents. Joe is writing a report on the backgrounds of American presidents. He looks up the ages of all 43 presidents when they entered office. Because Joe took a statistics course, he uses these 43 numbers to get a 95% confidence interval for the mean age of all men who have been president. This makes no sense. Why not?

6.5 How far do rich parents take us? How much education children get is strongly associated with the wealth and social status of their parents. In social science jargon, this is "socioeconomic status," or SES. But the SES of parents has little influence on whether children who have graduated from college go on to yet more education. One study looked at whether college graduates took the graduate admissions tests for business, law, and other graduate programs. The effects of the parents' SES on taking the LSAT test for law school were "both statistically insignificant and small."

Chapter 6   27

(a) What does "statistically insignificant" mean?
(b) Why is it important that the effects were small in size as well as insignificant?

6.6 Color blindness in Africa. An anthropologist suspects that color blindness is less common in societies that live by hunting and gathering than in settled agricultural societies. He tests a number of adults in two populations in Africa, one of each type. The proportion of color-blind people is significantly lower ($P < 0.05$) in the hunter-gatherer population. What additional information would you want to help you decide whether you accept the claim about color blindness?

6.7 Blood types in Southeast Asia. One way to assess whether two human groups should be considered separate populations is to compare their distributions of blood types. An anthropologist finds significantly different ($P = 0.01$) proportions of the main human blood types (A, B, AB, O) in different tribes in central Malaysia. What other information would you want before you agree that these tribes are separate populations?

6.8 Explaining statistical confidence. Here is an explanation from the Associated Press concerning one of its opinion polls. Explain briefly but clearly in what way this explanation is incorrect.

> *For a poll of 1,600 adults, the variation due to sampling error is no more than three percentage points either way. The error margin is said to be valid at the 95 percent confidence level. This means that, if the same questions were repeated in 20 polls, the results of at least 19 surveys would be within three percentage points of the results of this survey.*

6.9 California brush fires. We often see televised reports of brush fires threatening homes in California. Some people argue that the modern practice of quickly putting out small fires allows fuel to accumulate and so increases the damage done by large fires. A detailed study of historical data suggests that this is wrong—the damage has risen simply because there are more houses in risky areas.[21] As usual, the study report gives statistical information tersely. Here is the summary of a regression of number of fires on decade (9 data points, for the 1910s to the 1990s):

> *Collectively, since 1910, there has been a highly significant increase ($r^2 = 0.61$, $P < 0.01$) in the number of fires per decade.*

How would you explain this statement to someone who know no statistics? Include an explanation of both the description given by r^2 and its statistical significance.

6.10 Alcohol and mortality. It appears that people who drink alcohol in moderation have lower death rates than either people who drink heavily or people who do not drink at all. The protection offered by moderate drinking is concentrated among people over 50 and on deaths from heart disease. The Nurses Health Study played an essential role in establishing these facts for women. This part of the study followed 85,709 female nurses for 12 years, during which time 2658 of the subjects died. The nurses completed a questionnaire that described their diet, including their use of alcohol. They were reexamined every two years. Conclusion: "As compared with nondrinkers and heavy drinkers, light-to-moderate

drinkers had a significantly lower risk of death."[22]
(a) Was this study an experiment? Explain your answer.
(b) What does "significantly lower risk of death" mean in simple language?
(c) Suggest some lurking variables that might be confounded with how much a person drinks. The investigators used advanced statistical methods to adjust for many such variables before concluding that the moderate drinkers really have a lower risk of death.

6.11 Alcohol and mortality, continued. The Nurses Health Study (see the previous exercise) reported the *relative risk* of death for women who drink different amounts of alcohol. (A relative risk of 2 means people in this group were twice as likely as nondrinkers to die during the study period. The risks were adjusted to correct for other differences among the groups.) Here are the facts:

Group	Relative risk compared with nondrinkers	95% confidence interval for risk
Light	0.83	0.74 to 0.93
Moderate	0.88	0.80 to 0.98
Heavy	1.19	1.02 to 1.38

Roughly speaking, light drinkers report one to three drinks per week, moderate drinkers four to 20 drinks per week (without bingeing!) and heavy drinkers more than 20 drinks per week.
(a) Can you be quite confident that light drinkers have a lower risk of death than nondrinkers? Why?
(b) Can you be quite confident that light drinkers have a lower risk of death than moderate drinkers? Why?
(c) Can you be quite confident that heavy drinkers have a higher risk of death than nondrinkers? Why?

Confidence intervals applet exercises

The interactive applets for *The Basic Practice of Statistics* are found on the BPS companion Web site, www.whfreeman.com/bps. The confidence intervals applet animates Figure 6.4 on page 302 of BPS. This applet is the best way I know to grasp the idea of a confidence interval.

6.12 80% confidence intervals. The idea of a 80% confidence interval is that the interval captures the true parameter value in 80% of all samples. That's not high enough confidence for practical use, but 80% hits and 20% misses make it easy to see how a confidence interval behaves in repeated samples from the same population.
(a) Set the confidence level in the applet to 80%. Click "Sample" to choose an SRS and calculate the confidence interval. Do this 10 times to simulate 10 SRSs with their 10 confidence intervals. How many of the 10 intervals captured the true mean μ? How many missed?
(b) You see that we can't predict whether the next sample will hit or miss. The confidence level, however, tells us what percent will hit in the long run. Reset the applet and click "Sample 50" to get the confidence intervals from 50 SRSs. How many hit? Keep clicking

Chapter 6

"Sample 50" and record the percent of hits among 100, 200, 300, 400, 500, 600, 700, 800, and 1000 SRSs. Even 1000 samples is not truly "the long run," but we expect the percent of hits in 1000 samples to be fairly close to the confidence level, 80%.

6.13 What confidence means. Confidence tells us how often our method will produce an interval that captures the true population parameter if we use the method a very large number of times. The applet allows us to actually use the method many times.
(a) Set the confidence level in the applet to 90%. Click "Sample 50" to choose 50 SRSs and calculate the confidence intervals. How many captured the true population mean μ? Keep clicking "Sample 50" until you have 1000 samples. What percent of the 1000 confidence intervals hit?
(b) Now choose 95% confidence. Look carefully when you first click "Sample 50." Are these intervals longer or shorter than the 90% confidence intervals? Again take 1000 samples. What percent of the intervals captured the true μ?
(c) Do the same thing for 99% confidence. What percent of 1000 samples gave confidence intervals that caught the true mean? Did the behavior of many intervals for the three confidence levels closely reflect the choice of confidence level?

Test of significance applet exercises

The test of significance applet animates Example 6.7 on page 319 of BPS. That example asks if a basketball player's actual performance gives evidence against the claim that he or she makes 80% of free throws.

The parameter in question is the percent p of free throws that the player will make if he or she shoots free throws forever. The population is all free throws the player will ever shoot. The null hypothesis is always the same, that the player makes 80% of shots taken:

$$H_0 : p = 80\%$$

The applet does not do a formal statistical test. Instead, it allows you to ask the player to shoot until you are reasonably confident that the true percent of hits is or is not very close to 80%.

6.14 I'm a great free-throw shooter. I claim that I make 80% of my free throws. To test my claim, we go to the gym and I shoot 20 free throws. Set the applet to take 20 shots. Check "Show null hypothesis" so that my claim is visible in the graph.
(a) Click "Shoot." How many of the 20 shots did I make? Are you convinced that I really make less than 80%?
(b) If you are not convinced, click "Shoot" again for 20 more shots. Keep going until *either* you are convinced that I don't make 80% of my shots *or* it appears that my true percent made is pretty close to 80%. How many shots did you watch me shoot? How many did I make? What did you conclude? Then click "Show true %" to reveal the truth. Was your conclusion correct?

(*Comment*: You see why statistical tests say how strong is the evidence *against* some claim. If I make only 10 of 40 shots, you are pretty sure I can't make 80% in the long run. But even if I make exactly 80 of 100, my true long-term percent might be 78% or 81% instead of 80%. It's hard to be convinced that I make exactly 80%.)

6.15 Maybe she's better (or worse). Clicking "New shooter" in the applet brings a new player to the free-throw line. We don't know how good this player is. In the previous exercise, you decided either that I don't make 80% or that my true percent is at least pretty close to 80%. Now use the applet just as in the last exercise until you come upon a player whose result is different from mine. You may have to try several new shooters. (Be honest: Make the player shoot to see data before you click "Show true %" to see what lies behind the data.) If you decided that I make 80%, find a shooter who makes less than 80%. If you decided that I don't make 80%, find a shooter who makes very close to 80%. For each shooter until you find what you are looking for, record how many shots you watched, what percent were made, what you concluded about this shooter, and what "Show true %" told you about the player.

Normal curve applet exercises

6.16 The bottom row of Table C in BPS shows critical values for the standard normal distribution. For each of a number of upper-tail probabilities p, the table gives the value z of a standard normal variable having area p to its right under the normal curve. Use the applet to verify that the critical value for tail probability $p = 0.025$ is $z = 1.96$. Make a sketch of the curve from the applet marked with the values of p and z.

6.17 For reasons we can't guess, Max is interested in whether a one-sided z test is statistically significant at the $\alpha = 0.0125$ level. Use the applet to tell Max what values of z are significant. Sketch the standard normal curve marked with the values that led to your result.

6.18 Jaran wants to know if a two-sided z test statistic is significant at the $\alpha = 0.011$ level. Use the applet to say what values of z are significant at this level. Sketch the standard normal curve marked with the values that led to your result.

6.19 What standard normal critical value z^* is required for a 92.5% confidence interval for a population mean? Give the 92.5% confidence interval for the population mean in the following setting. The yield (bushels per acre) of variety of corn has standard deviation $\sigma = 10$ bushels per acre. Fifteen plots have these yields:

138.0 139.1 113.0 132.5 140.7 109.7 118.9 134.8
109.6 127.3 115.6 130.4 130.2 111.7 105.5

6.20 A one-sided z test for the hypotheses

$$H_0: \mu = 128$$
$$H_a: \mu < 128$$

has $z = -1.60$. Is this result statistically significant at the $\alpha = 0.055$ level? Explain with a sketch how you got your result.

CHAPTER 7 EXERCISES

7.1 Coaching and SAT scores. Coaching companies claim that their courses can raise the SAT scores of high school students. Of course, students who re-take the SAT without paying for coaching generally raise their scores. A random sample of students who took the SAT twice found 427 who were coached and 2,733 who were uncoached.[23] Starting with their Verbal scores on the first and second tries, we have these summary statistics:

	Try 1		Try 2		Gain	
	Mean	Std Dev	Mean	Std Dev	Mean	Std Dev
Coached	500	92	529	97	29	59
Uncoached	506	101	527	101	21	52

Let's first ask if students who are coached significantly increased their score.
(a) You could use the information given to carry out either a two-sample t test comparing Try 1 with Try 2 for coached students or a paired-sample t test using Gain. Which is the correct test? Why?
(b) Carry out the proper test. What do you conclude?
(c) Give a 99% confidence interval for the mean gain of all students who are coached.

7.2 Coaching and SAT scores, continued. What we really want to know is whether coached students improve more than uncoached students, and whether any advantage is large enough to be worth paying for. Use the information in the previous problem to answer these questions.
(a) Is there good evidence that coached students gained more on the average than uncoached students?
(b) How much more do coached students gain on the average? Give a 99% confidence interval.
(c) Based on your work, what is your opinion: Do you think coaching courses are worth paying for?

7.3 Coaching and SAT scores: critique. The data you used in the previous two problems came from a random sample of students who took the SAT twice. The response rate was 63%, which is pretty good for non-government surveys, so let's accept that the respondents do represent all students who took the exam twice. Nonetheless, we can't be sure that coaching actually *caused* the coached students to gain more than the uncoached students. Explain briefly but clearly why this is so.

7.4 A big toe problem. Hallux abducto valgus (call it HAV) is a deformation of the big toe that is not common in youth and often requires surgery. Doctors used X-rays to measure the angle (in degrees) of deformity in 38 consecutive patients under the age of 21 who came to a medical center for surgery to correct HAV. The angle is a measure of the seriousness of the deformity. Here are the data:[24]

```
28  32  25  34  38  26  25  18  30  26  28  13  20
21  17  16  21  23  14  32  25  21  22  20  18  26
16  30  30  20  50  25  26  28  31  38  32  21
```

We are willing to consider these patients as a random sample of young patients who require HAV surgery. Give a 95% confidence interval for the mean HAV angle in the population of all such patients.

7.5 A big toe problem, continued. The data in the previous problem follow a normal distribution quite closely except for one patient with HAV angle 50 degrees, a high outlier.
(a) Find the 95% confidence interval for the population mean based on the 37 patients who remain after you drop the outlier.
(b) Compare your interval in (a) with your interval from the previous problem. What is the most important effect of removing the outlier?

7.6 Significance. You are testing $H_0 : \mu = 0$ against $H_a : \mu \neq 0$ based on an SRS of 20 observations from a normal population. What values of the t statistic are statistically significant at the $\alpha = 0.005$ level?

7.7 What critical value? You have an SRS of 15 observations from a normally distributed population. What critical value would you use to obtain a 98% confidence interval for the mean μ of the population?

7.8 Exercising to lose weight. Exercise 3.4 describes a study that compares the effects of several types of exercise. Here is a small part of the summary of the results:

> *There was no significant difference between the LB and SB groups for mean (SD) weight loss at 18 months (LB, -5.8 [7.1] kg; SB, -3.7 [6.6] kg.)*

That's pretty terse, but it gives the mean and standard deviation of the weight loss (in kilograms) for the long bout (LB) and short bout (SB) groups. The data come from the 37 LB subjects and 36 SB subjects who completed the study. Do an appropriate test to confirm the report that there is not a significant difference in the mean weight loss in the two groups.

7.9 Way down under. A remarkable discovery: Large lakes exist deep under the antarctic ice cap, kept liquid by the enormous pressure of the ice above. The largest is Lake Vostok. Do these lakes contain populations of ancient bacteria adapted to the dark, cold, high-pressure environment? Drilling over 3600 meters (over 11,000 feet) down to a depth where the ice consists of water frozen from Lake Vostok did indeed find bacteria. The researchers estimated "mean bacterial biomass" in nanograms of carbon per liter of melted water. They had 5 specimen ice cores. They split each specimen into "top" and "bottom" to get two samples of size 5, which they processed separately. Separate study of the two samples is a check on the complicated process of preparing the cores for analysis. The results "are mean estimates ±SD." Here is an extract:[25]

	Top melt	Bottom melt
Mean bacterial biomass (ng of C liter^{-1})	2.9 ± 0.4	2.8 ± 0.4

Use the results for the 5 top melt specimens to give a 90% confidence interval for the mean bacterial biomass in the ice above Lake Vostok.

7.10 Way down under, continued. Do the results reported in the previous exercise give any reason to think that the population mean biomasses as estimated from "top melt" and "bottom melt" samples are different? (A difference would suggest that the preparation process influences the findings.)

7.11 Mutual fund performance. Exercise 2.8 gives data on the annual returns (percent) for the Vanguard International Growth Fund and its benchmark index, the Morgan Stanley EAFE index. (The data set is `E02-08.dat`.) Does the fund significantly outperform its benchmark?
(a) Explain clearly why the matched pairs t test, *not* the two-sample t test, is the proper choice to answer this question.
(b) Make a stemplot of the differences (fund − EAFE) for the 19 years. There is no reason to doubt approximate normality of the differences. (More detailed study shows that the differences follow a normal distribution quite closely.)
(c) Carry out the test and state your conclusion about the fund's performance.

7.12 Growing trees faster. The concentration of carbon dioxide (CO_2) in the atmosphere is increasing rapidly due to our use of fossil fuels. Because plants use CO_2 to fuel photosynthesis, more CO_2 may cause trees and other plants to grow faster. An elaborate apparatus allows researchers to pipe extra CO_2 to a 30-meter circle of forest. They selected two nearby circles in each of three parts of a pine forest and randomly chose one of each pair to receive extra CO_2. The response variable is the mean increase in base area for 30 to 40 trees in a circle during a growing season. We measure this in percent increase per year. Here are one year's data:[26]

Pair	Control plot	Treated plot
1	9.752	10.587
2	7.263	9.244
3	5.742	8.675

(a) State the null and alternative hypotheses. Explain clearly why the investigators used a one-sided alternative.
(b) Carry out a test and report your conclusion in simple language.

7.13 Mouse genes. A study of genetic influences on diabetes compared normal mice with similar mice genetically altered to remove the gene called $aP2$. Mice of both types were allowed to become obese by eating a high-fat diet. The researchers then measured the levels of insulin and glucose in their blood plasma. Here are some excerpts from their findings.[27] The normal mice are called "wild-type" and the altered mice are called "$aP2^{-/-}$."

> Each value is the mean ± SEM of measurements on at least 10 mice. Mean values of each plasma component are compared between $aP2^{-/-}$ mice and wild-type controls by Student's t test (*$P < 0.05$ and **$P < 0.005$).
>
> | Parameter | Wild type | $aP2^{-/-}$ |
> |------------------|------------|----------------|
> | Insulin (ng/ml) | 5.9 ± 0.9 | 0.75 ± 0.2** |
> | Glucose (mg/dl) | 230 ± 25 | 150 ± 17* |
>
> Despite much greater circulating amounts of insulin, the wild-type mice had higher blood glucose than the $aP2^{-/-}$ animals. These results indicate that the

absence of aP2 interferes with the development of dietary obesity-induced insulin resistance.

Other biologists are supposed to understand the statistics reported so tersely.
(a) What does "SEM" mean? What is the expression for SEM based on n, $\bar{x}$, and s from a sample?
(b) Which of the tests we have studied did the researchers apply?
(c) Explain to a biologist who knows no statistics what $P < 0.05$ and $P < 0.005$ mean. Which is stronger evidence of a difference between the two types of mice?

7.14 Mouse genes, continued. The report quoted in the previous exercise says only that the sample sizes were "at least 10." Suppose that the results are based on exactly 10 mice of each type. Use the values in the table to find $\bar{x}$ and s for the insulin concentrations in the two types of mice. Carry out a test to assess the significance of the difference in mean insulin concentration. Does your P-value confirm the claim in the report that $P < 0.005$?

Use of applets in Chapters 7 to 12

Chapters 7 to 12 of *The Basic Practice of Statistics* concern methods for statistical inference (tests and confidence intervals) in a variety of settings. Statistical software and some advanced calculators will carry out the calculations for inference. The interactive applets found on the BPS companion Web site, `www.whfreeman.com/bps`, do not as yet include applets that carry out inference procedures or help you understand them. If you do not use software, you will nonetheless find the applets useful for graphing your data before inference and for basic calculations. If, for example, you are considering use of the one-sample t procedures, you will want to make a histogram to check for outliers or extreme skewness and then calculate $\bar{x}$ and s. The one-variable statistical calculator automates these steps. Watch the Web site for new applets on such topics as analysis of variance.

CHAPTER 8 EXERCISES

8.1 Spinning pennies. We suspect that spinning a penny, unlike tossing it, does not give heads and tails equal probabilities. I spun a penny 200 times and got 83 heads. How significant is this evidence against equal probabilities?

8.2 The millennium begins with optimism. In January of the year 2000, the Gallup Poll asked a random sample of 1633 adults, "In general, are you satisfied or dissatisfied with the way things are going in the United States at this time?" It found that 1127 said that they were satisfied. Write a short report of this finding, as if you were writing for a newspaper. Be sure to include a margin of error.

8.3 Heads in spinning pennies. I once saw a statistics student sitting in a corner at a party spinning pennies and betting on tails at even odds. The people betting against him thought heads had probability 0.5. Use the results of 100 spins in Exercise 8.1 to give a 90% confidence interval for the true probability of a head in spinning pennies.

8.4 How common is SAT coaching? A random sample of students who took the SAT college entrance examination twice found that 427 of the respondents had paid for coaching courses and that the remaining 2,733 had not.[28] Give a 99% confidence interval for the proportion of coaching among students who re-take the SAT.

8.5 How to quit smoking. Nicotine patches are often used to help smokers quit. Does giving medicine to fight depression help? A randomized double-blind experiment assigned 244 smokers who wanted to stop to receive nicotine patchs and another 245 to receive both a patch and the anti-depression drug bupropion. Results: After a year, 40 subjects in the nicotine patch group had abstained from smoking, as had 87 in the patch-plus-drug group.[29] Is this good evidence that adding bupropion increases the success rate? (State hypotheses, calculate a test statistic, and give a P-value and your conclusion.)

8.6 Genetically modified foods. Europeans have been more skeptical than Americans about the use of genetic engineering to improve foods. A sample survey gathered responses from random samples of 863 Americans and 12,178 Europeans.[30] (The European sample was larger because Europe is divided into several nations.) Subjects were asked to consider

> *Using modern biotechnology in the production of foods, for example to make them higher in protein, keep longer, or change in taste.*

They were asked if they considered this "risky for society." In all, 52% of Americans and 64% of Europeans thought the application was risky.
(a) It is clear without a formal test that the proportion of the population who consider this use of technology risky is significantly higher in Europe than in the United States. Why is this?
(b) Give a 99% confidence interval for the percentage difference between Europe and the United States.

8.7 Genetically modified foods, continued. Is there convincing evidence that more than half of all adult Americans consider applying biotechnology to the production of foods risky?

8.8 Genetically modified foods, continued. Give a 95% confidence interval for the proportion of all European adults who consider the use of biotechnology in food production risky.

8.9 Satisfaction with high schools. A sample survey asked 202 black parents and 201 white parents of high-school children, "Are the public high schools in your state doing an excellent, good, fair or poor job, or don't you know enough to say?" The investigators suspected that black parents are generally less satisfied with their public schools than are whites. Among the black parents, 81 thought high schools were doing a "good" or "excellent" job; 103 of the white parents felt this way.[31] Is there good evidence that the proportion of all black parents who think their state's high schools are good or excellent is lower than the proportion of white parents with this opinion?

8.10 College is important. The sample survey described in the previous exercise also asked respondents if they agreed with the statement, "A college education has become as important as a high school diploma used to be." In the sample, 125 of 201 white parents and 154 of 202 black parents said that they "strongly agreed." Is there good reason to think that different percents of all black and white parents would strongly agree with the statement?

8.11 Regulation of guns. The 1998 National Gun Policy Survey, conducted by the National Opinion Research Center at the University of Chicago, asked many questions about regulation of guns in the United States. The sample can be considered an SRS of adult residents of the United States. It was conducted by calling randomly selected telephone numbers in all 50 states. When a household answered the phone, the questions were asked of the adult with the most recent birthday. The response rate was 60.5%. In all, 1204 people responded to the survey.

One of the questions was "Do you think there should be a law that would ban possession of handguns except for the police and other authorized persons?" Here are some results for this question, based on the 1201 people who gave both their gender and their level of education:[32]

	Men		Women
Yes	148	Yes	338
No	383	No	332

Give a confidence interval for the proportion of all adults who favor a ban on personal possession of handguns.

8.12 Regulation of guns, continued. Women generally support restrictions on firearms more strongly than men. Do the data in the previous exercise provide good evidence that the proportion of women who favor banning possession of handguns is greater than the proportion of men who favor a ban?

CHAPTER 9 EXERCISES

9.1 How to quit smoking. It's hard for smokers to quit. Perhaps prescribing a drug to fight depression will work as well as the usual nicotine patch. Perhaps combining the patch and the drug will work better than either alone. Here are data from a randomized, double-blind trial that compared four treatments.[33] A "success" means that the subject did not smoke for a year following the beginning of the study.

Treatment	Subjects	Successes
Nicotine patch	244	40
Drug	244	74
Patch plus drug	245	87
Placebo	160	25

(a) Make a graph that compares the rates of success for the four treatments.
(b) Make a two-way table of successes and failures for the four treatments. Is there good evidence that the proportion of success for all smokers differs among the treatments? (State hypotheses, give a test statistic and its P-value, and state your conclusion.)

9.2 Do fruit flies sleep? Mammals and birds sleep. Fruit flies show a daily cycle of rest and activity, but does the rest qualify as sleep? Researchers looking at brain activity and behavior finally concluded that fruit flies do sleep. A small part of the study used an infra-red motion sensor to see if flies moved in response to vibrations. Here are results for low levels of vibration:[34]

	Response to vibration?	
	No	Yes
Fly was walking	10	54
Fly was resting	28	4

Analyze these results. Is there good reason to think that resting flies respond differently than flies that are walking? (That's a sign that the resting flies may actually be sleeping.)

9.3 Shaking fruit flies. In planning the experiment described in the previous exercise, the researchers chose several levels of vibration. We will call these just "Low," "Medium," and "High." Here are the responses of flies that were walking to these three levels of stimulus.

	Response?	
	No	Yes
Low	10	54
Medium	3	52
High	0	36

What percent of the sample flies responded to each level of vibration? Do you think that the percentage of all fruit flies that would respond changes with the level of vibration?

9.4 Dropouts. It is common for some subjects to drop out of experiments before the study is complete. If the dropout rate differs greatly among the treatments, we worry that dropping out or not is related to the treatments given. If that is so, we may not be able to trust analyses based on the subjects who stick it out. Investigators therefore always look at dropout rates. Here are the dropout data from the weight loss study described in detail in Exercise 3.4:

	Completed 18 months?	
Treatment	Yes	No
Long-bout exercise	37	12
Short-bout exercise	36	15
Short-bout with equipment	42	6

(a) What are the percentages of dropouts in each group?
(b) State carefully the hypotheses for the chi-square test in this setting. (See Exercise 3.4 to describe the population.) Carry out the test and explain what it tells you.

9.5 How are schools doing? The nonprofit group Public Agenda conducted telephone interviews with randomly selected parents of high-school children. There were 202 black parents, 202 Hispanic parents, and 201 white parents. One question asked was, "Are the high schools in your state doing an excellent, good, fair or poor job, or don't you know enough to say?" Here are the survey results:[35]

	Black parents	Hispanic parents	White parents
Excellent	12	34	22
Good	69	55	81
Fair	75	61	60
Poor	24	24	24
Don't know	22	28	14
TOTAL	202	202	201

Write a brief analysis of these results. Include a graph or graphs, a test of significance, and your own discussion of the most important findings.

9.6 College is important. The sample survey described in the previous exercise also asked the parents to respond to the statement, "A college education has become as important as a high school diploma used to be." Here are the counts of responses:

	Black parents	Hispanic parents	White parents
Strongly agree	154	144	125
Somewhat agree	27	37	50
Somewhat disagree	11	13	18
Strongly disagree	10	8	8
TOTAL	202	202	201

Write a brief report on the similarities or differences among the three groups of parents in their attitudes towards the importance of a college education. Include a graph or graphs and a test of significance.

9.7 Regulating guns. Exercise 8.11 describes the National Gun Policy Survey. One of the questions asked was "Do you think there should be a law that would ban possession of handguns except for the police and other authorized persons?" Here are the responses, broken down by the respondent's level of education:

	Yes	No
Less than high school	58	58
High school graduate	84	129
Some college	169	294
College graduate	98	135
Postgraduate degree	77	99

(a) How does the proportion of the sample who favor banning possession of handguns differ among people with different levels of education? Make a bar graph that compares the proportions and briefly describe the relationship between education and opinion about a handgun ban.

(b) Does the sample provide good evidence that the proportion of the adult population who favor a ban on handguns changes with level of education?

9.8 Early to bed? Is it true that, "Early to bed and early to rise makes a man healthy, wealthy, and wise?" A study of older people in England suggests that Benjamin Franklin's saying no longer applies. The subjects were 1229 randomly selected adults who were at least 65 years old in 1973. The subjects were followed for 23 years to look at such things as mortality and cause of death.

The investigators call a subject an "owl" if they regularly go to bed after 11 PM and rise at or after 8 AM. A subject is a "lark" if they retire before 11 PM and rise before 8 AM. The overall conclusion was that owls actually do a bit better than larks in many respects. Here is a two-way table for one response variable, access to a car at the start of the study in 1973:[36]

	Access to car?	
	Yes	No
Larks	122	234
Owls	138	180
Other sleeping patterns	213	342

Write a brief report of the relationship between sleeping pattern and access to a car. Include a graph and a test of significance.

CHAPTER 10 EXERCISES

10.1 Do fruit flies sleep? Mammals and birds sleep. Insects such as fruit flies rest, but is this rest sleep? Investigators looked at many different aspects of rest in fruit flies and concluded that the insects do indeed sleep. One experiment in the study gave caffeine to fruit flies to see if it affects their rest. We know that caffeine reduces sleep in mammals, so if it reduces rest in fruit flies that's another hint that the rest is really sleep. The paper reporting the study contains a graph similar to Figure 4 and states that, "Flies given caffeine obtained less rest during the dark period in a dose-dependent fashion ($n = 36$ per group, $P < 0.0001$)."[37]

(a) The explanatory variable is amount of caffeine, in milligrams per milliliter of blood. The response variable is minutes of rest (measured by an infra-red motion sensor) during a 12-hour dark period. Outline the design of this experiment.

(b) The P-value in the report comes from the ANOVA F-test. State in words the null and alternative hypotheses for the test in this setting. What do the graph and the statistical test together lead you to conclude?

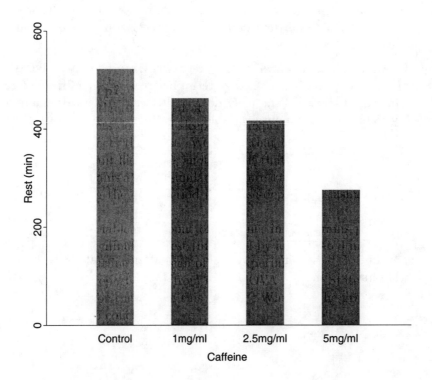

Figure 4: Mean rest time for fruit flies given several doses of caffeine.

10.2 Exercising to lose weight. Exercise 3.4 describes a randomized, comparative experiment that assigned subjects to three types of exercise program intended to help them lose weight. Some of the results of this study were analyzed using the chi-square test for two-way tables, and some others were analyzed using one-way ANOVA. For each of the following excerpts from the study report, say which analysis is appropriate and explain how you made your choice.

(a) "Overall, 115 subjects (78% of 148 subjects randomized) completed 18 months of treatment, with no significant difference in attrition rates between the groups ($P = .12$)."

(b) "In analyses using only the 115 subjects who completed 18 months of treatment, there were no significant differences in weight loss at 6 months among the groups."

(c) "The duration of exercise for weeks 1 through 4 was significantly greater in the SB compared with both LB and SBEQ groups ($P < .05$). ... However, exercise duration was greater in SBEQ compared with both LB and SB groups for months 13 through 18 ($P < .05$)."

10.3 Plants defend themselves. When some plants are attacked by leaf-eating insects, they release chemical compounds that attract other insects that prey on the leaf-eaters. A study carried out on plants growing naturally in the Utah desert demonstrated both the release of the compounds and that they not only repel the leaf-eaters but attract predators that act as the plant's bodyguards.[38] The investigators chose 8 plants attacked by each of three leaf-eaters and 8 more that were undamaged, 32 plants of the same species in all. They then measured emissions of several compounds during seven hours. Here are data (mean ± SEM for eight plants) for one compound. The emission rate is measured in nanograms (ng) per hour.

Group	Emission rate, ng per hour
Control	9.22 ± 5.93
Hornworm	31.03 ± 8.75
Leaf bug	18.97 ± 6.64
Flea beetle	27.12 ± 8.62

(a) Make a graph that compares the mean emission rates for the four groups. Does it appear that emissions increase when the plant is attacked?

(b) What hypotheses does ANOVA test in this setting?

(c) We do not have the full data. What would you look for in deciding whether you can safely use ANOVA?

(d) What is the relationship between the standard error of the mean (SEM) and the standard deviation for a sample? Do the standard deviations satisfy our rule of thumb for safe use of ANOVA?

10.4 ANOVA details: weight loss. The calculations of ANOVA use only the sample sizes n_i, the sample means $\bar{x}_i$, and the sample standard deviations s_i. You can therefore recreate the ANOVA calculations when a report gives these summaries but does not give the actual data. The report for the weight-loss experiment described in Exercise 3.4 contains the following information about weight loss (in kilograms) after six months of treatment:

Treatment	n	$\bar{x}$	s
Long-bout exercise	37	10.2	4.2
Short-bout exercise	36	9.3	4.5
Short-bout with equipment	42	10.2	5.2

(a) Do the standard deviations satisfy the rule of thumb for safe use of ANOVA?

(b) Calculate the overall mean response $\bar{x}$ and the mean square for groups MSG.

(c) Calculate the mean square for error, MSE.

(d) Find the ANOVA F statistic and its approximate P-value. Is there evidence that the mean weight losses of people who follow the three exercise programs differ?

10.5 ANOVA details: plant defenses. The calculations of ANOVA use only the sample sizes n_i, the sample means $\bar{x}_i$, and the sample standard deviations s_i. You can therefore recreate the ANOVA calculations when a report gives these summaries but does not give the actual data. Use the information in Exercise 10.3 to calculate the ANOVA table (sums of squares, degrees of freedom, mean squares, and the F statistic). Note that the report gives the standard error of the mean (SEM) rather than the standard deviation. Are there significant differences among the mean emission rates for the four populations of plants?

10.6 Earnings of athletic trainers. How much do newly-hired athletic trainers earn? Earnings depend on the educational and other qualifications of the trainer and on the type of job. Here are summaries from a sample survey that contacted institutions advertising for trainers and asked about the people they hired.[39]

Type of institution	n	Mean salary $\bar{x}$	Std dev s
High school	57	$26,470	$9507
Clinic	108	$30,610	$4504
College	94	$30,019	$7158

Calculate the ANOVA table and the F statistic. Are there significant differences among the mean salaries for the three types of job in the population of all newly-hired trainers? (*Comments*: The study authors did an ANOVA. As is often the case, the published work does not answer questions about the suitability of the analysis. We wonder if responses from 60% of employers who advertised positions generate a random sample of new hires. The actual data do not appear in the report. The salary distributions are no doubt right-skewed; we hope that strong skewness and outliers are absent because the subjects in each group hold similar jobs. The large samples will then justify ANOVA. Similarly, the sample standard deviations do not quite satisfy our rule of thumb for safe use of ANOVA.)

CHAPTER 11 EXERCISES

11.1 Age and income. Figure 2 (Exercise 2.1) is a scatterplot of the age and income of a random sample of 5712 men between the ages of 25 and 65 who have a bachelor's degree but no higher degree. We see that estimated mean income does increase with age, but not very rapidly.
(a) We know even without looking at the software output in Exercise 2.1 that there is highly significant evidence that the slope of the true regression line is greater than 0. Why do we know this?
(b) The software gives the two-sided P-value for a test of $H_0 : \beta = 0$. Before collecting the data, we suspected that (on the average) older men earn more. Our alternative hypothesis is therefore $H_a : \beta > 0$. What is the P-value for this one-sided alternative?

11.2 Age and income, continued. Give a 99% confidence interval for the slope of the true regression line of income on age for all men aged 25 to 65 with a bachelor's degree but no higher degree. Use the information in the software output in Exercise 2.1.

11.3 Stocks and bonds. Exercise 2.6 presents data on the relationship between the flows of investment dollars into stock mutual funds and bond mutual funds. The descriptive analysis in that exercise shows a negative relationship—bond investment tends to go up when stock investment goes down. (The data set is E02-06.dat.)
(a) Is there statistically significant evidence that there is some straight-line relationship between the flows of cash into bond funds and stock funds? (State hypotheses, give a test statistic and its P-value, and state your conclusion.)
(b) What fact about your scatterplot in Exercise 2.6 explains why the relationship described by the least-squares line is not significant?

11.4 Foot problems. Exercise 2.2 and Table 2 describe the relationship between two deformities of the feet in young patients. Metatarsus adductus may help predict the severity of hallux abducto valgus. (The data set is E02-02.dat.) The paper that reports this study says, "Linear regression analysis, using the hallux abductus angle as the response variable, demonstrated a significant correlation between the metatarsus adductus and hallux abductus angles."[40] Do a suitable test of significance to verify this finding. (The study authors then note that the scatterplot suggests that the variation in y may change as x changes and offer a more elaborate analysis as well.)

SOURCES

1. From the National Association of Realtors Web site, `nar.realtor.com`.

2. Alan S. Banks et al., "Juvenile hallux abducto valgus association with metatarsus adductus," *Journal of the American Podiatric Medical Association* 84 (1994), pp. 219–224.

3. Data from the Web site of Professor Kenneth French of MIT, `web.mit.edu/kfrench/www/data_library.html`.

4. J. Marcus Jobe and Hutch Jobe, "A statistical approach for additional infill development," *Energy Exploration and Exploitation*, 18 (2000), pp. 89–103.

5. Reed Abelson, "Companies turn to grades, and employees go to court," *New York Times*, March 19, 2000.

6. Data from the March, 2000 Annual Demographic Supplement to the Current Population Survey. These data were obtained from the Web site of the Bureau of Labor Statistics, `stats.bls.gov/datahome.htm`.

7. See Note 2.

8. "Dancing in step," *Economist*, March 22, 2001.

9. Net cash flow data from Sean Collins, *Mutual Fund Assets and Flows in 2000*, Investment Company Institute, 2001. Found online at `www.ici.org`. The raw data were converted to real dollars using annual average values of the CPI.

10. Illiteracy data from the Social Indicators Home Page of the United Nations Statistical Division, `www.un.org/Depts/unsd/`.

11. From the performance data for the fund presented at the Vanguard Group Web site, `personal.vanguard.com`.

12. Carol A. Warfield, "Controlled-release morphine tablets in patients with chronic cancer pain," *Cancer*, 82 (1998), pp. 2299–2306.

13. E. M. Peters et al., "Vitamin C supplementation reduces the incidence of postrace symptoms of upper-respiratory tract infection in ultramarathon runners," *American Journal of Clinical Nutrition*, 57 (1993), pp. 170–174.

14. John M. Jakicic et al., "Effects of intermittent exercise and use of home exercise equipment on adherence, weight loss, and fitness in overweight women," *Journal of the American Medical Association*, 282 (1999), pp. 1554–1560.

15. Susan B. Sorenson, "Regulating firearms as a consumer product," *Science*, 286 (1999), pp. 1481–1482, Note 12.

16. From *CIS Boletin 9, Spaniards' Economic Awareness*, found online at `cis.sociol.es/ingles`.

17. NIMH Multisite HIV Prevention Trial Group, "The NIMH multisite HIV prevention trial: reducing HIV sexual risk behavior," *Science*, 280 (1998), pp. 1889–1894.

Sources

18. Based on Evan H. DeLucia et al., "Net primary production of a forest ecosystem with experimental CO_2 enhancement," *Science*, 284 (1999), pp. 1177–1179. The investigators used the block design.

19. Based on the Electronic Encyclopedia of Statistical Examples and Exercises (EESEE) story "Surgery in a blanket," found on the BPS Web site www.whfreeman.com/bps.

20. Warren E. Leary, "Cell phones: questions but no answers," *New York Times*, October 26, 1999.

21. Jon E. Keeley, C. J. Fotheringham, and Marco Morais, "Reexamining fire suppression impacts on brushland fire regimes," *Science*, 284 (1999), pp. 1829–1831.

22. Charles S. Fuchs, et al., "Alcohol consumption and mortality among women," *New England Journal of Medicine*, 332 (1995), pp. 1245–1250.

23. Wayne J. Camera and Donald Powers, "Coaching and the SAT I," *TIP* (online journal: www.siop.org/tip), July 1999.

24. See Note 2.

25. D. M. Karl et al., "Microorganisms in the accreted ice of Lake Vostok, Antarctica," *Science*, 286 (1999), pp. 2144–2147.

26. See Note 18. I thank the authors for supplying the actual data for this part of their study.

27. Gökhan S. Hotamisligil et al., "Uncoupling of obesity from insulin resistance through a targeted mutation in $aP2$, the adipocyte fatty acid binding protein," *Science*, 274 (1996), pp. 1377–1379.

28. See Note 23.

29. Douglas E. Jorenby et al., "A controlled trial of sustained-release bupropion, a nicotine patch, or both for smoking cessation," *New England Journal of Medicine*, 340 (1999), pp. 685–691.

30. Based on the online supplement to G. Gaskell et al., "Worlds apart? The reception of genetically modified foods in Europe and the U.S.," *Science*, 285 (1999), pp. 384–387. The percents given are not exact because of rounding in the table from which they are compiled.

31. Data compiled from a table of percents in "Americans view higher education as key to the American dream," press release by The National Center for Public Policy and Higher Education, www.highereducation.org, May 3, 2000.

32. Based closely on Susan B. Sorenson, "Regulating firearms as a consumer product," *Science*, 286 (1999), pp. 1481–1482. Because the results in the paper were "weighted to the U.S. population," I have changed some counts slightly for consistency.

33. See Note 29.

34. Based on the online supplement to Paul J. Shaw et al., "Correlates of sleep and waking in *Drosophila melanogaster*," *Science*, 287 (2000), pp. 1834–1837.

35. See Note 31.

36. Catharine Gale and Christopher Martyn, "Larks and owls and health, wealth, and wisdom," *British Medical Journal*, 317 (1998), pp. 1675–1677.

37. See Note 34.

38. Data from the online supplement to André Kessler and Ian T. Baldwin, "Defensive function of herbivore-induced plant volatile emissions in nature," *Science*, 291 (2001), pp. 2141–2144.

39. Based on Brent L. Arnold et al., "1994 athletic trainer employment and salary characteristics," *Journal of Athletic Training*, 31 (1996), pp. 215–218. I have updated the 1994 salaries to the equivalent in 2000 dollars, using the Consumer Price Index, to avoid misleadingly low values.

40. See Note 2.

ANSWERS TO
ODD-NUMBERED ADDITIONAL EXERCISES

Note to students: These answers are intended to serve as "benchmarks" with which you can compare your results; they should not be considered models of ideal responses to these exercises. They typically do not show the work required to perform a statistical test or construct a confidence interval, and rarely are written in complete sentences. Also missing are graphs, stemplots, experimental design diagrams, etc. Your responses should typically be written so that someone who has not read the question can make sense of your answer.

Darryl K. Nester
Bluffton College

CHAPTER 1

1.1 (a) Mutual funds. (b) Four: category and largest holding (categorical), net assets and YTD return (quantitative). (c) Net assets in $million; YTD return is given as a percentage.

1.3 Use either a stemplot or a histogram. Five-number summary (in degrees): 13, 20, 25, 30, 50. Fairly symmetric, one high outlier, most angles ranging from 13 to 38 degrees, and half between 20 and 30 degrees. The center is about 25 degrees.

1.5 (a) Reasonably symmetric except for the outlier. (b) Center: slightly more than 0. (c) Smallest: between -16% and -14%. Largest: between 16% and 18%. (d) About 37%.

1.7 (a) Looking at number of doctors per 100,000 people allows comparison of states with different populations. (b) Distribution is right-skewed; D.C. is an outlier, presumably because it is more urban.

1.9 (a) Strongly right-skewed; the two highest values might be considered outliers. (b) $\bar{x} = 48.25$ thousand barrels; $M = 37.80$ thousand barrels; $\bar{x}$ is pulled in the direction of the skew. (c) In thousands of barrels: 2, 21.505, 37.8, 60.1, 204.9. Skewness causes the gap between the last two numbers.

1.11 About 7.8%.

1.13 About 1%.

1.15 (a) 31 classes. (b) One class. (c) 63 classes; 5 observations.

1.17 (a) The mean moves in the same direction as the moving point, while the median points to the right-most non-moving point. (b) The mean follows the moving point. As the moving point crosses the other two, the median slides along with it, then stays at the left-most fixed point.

1.19 (a) 0.6826; the 68–95–99.7 rule gives 0.68. (b) 0.9544 (compared to 0.95); 0.9974 (compared to 0.997).

1.21 A/B cutoff: mean plus 1.28 standard deviations; B/C cutoff: mean minus 1.28 standard deviations.

1.23 (a) About 3% of men score 750 or more. (b) About 1% of women score 750 or more.

CHAPTER 2

2.1 (a) Age is recorded as a whole number of years (no fractions). (b) E.g.: Older men have more experience, and more opportunities for promotion. Younger men might have more time and energy to devote to their jobs. The data shows that older men typically earn more, but the relationship is not very strong. (c) $\hat{y} \doteq 24,874 + 892.11x$. Salary rises an average of about $892 every year.

2.3 (a) $\hat{y} = 19.7 + 0.339x$. (b) About 28.2°. (c) A straight-line relationship with MA angle only explains about 9.1% of the variation in HAV angle.

2.5 Incorrect; it should be $r^2 = 64\%$.

2.7 (a) A positive linear association, with more scatter in the middle and upper right portions of the graph. $r \doteq 0.9350$. (b) $\hat{y} = 2.934 + 1.41x$. Largest residuals: Yemen (positive) and Lesotho (negative). Highest rates: Niger and Burkina Faso.

2.9 See answers to Exercise 2.7 above. To get names of countries for part (b), refer to the data file (they are not shown in the applet).

2.11 (a) $r = 1$ for a line. (c) Leave some space above your vertical stack. (d) The curve must be higher at the right than at the left.

2.13 (a) r will be closer to 1. (b) Results will vary.

2.15 (a) The green region indicates how far your line is from the least-squares line.

CHAPTER 3

3.1 (a) 29,777. (b) It is a voluntary response sample. (c) Men are likely to be over-represented in this sample.

3.3 (a) A block design.

3.5 (a) Dialing numbers randomly, rather than choosing from the phone book, allows access to all households with telephones. (b) The opinions of habitual phone-answerers might differ from those of the population as a whole.

3.7 (a) The control treatment, with only a one-hour session, also had a substantial increase. (b) Use a randomized comparative design. (c) Labels 0001–3706; choose 1887, 2099, 3547, 0426, 3543.

3.9 (a) There may be some difference between the two operating teams. (b) Randomly assign n patients to each group. (c) (Double) blindness.

3.11 Population = 1 to 30, sample size 15, then click "Reset" and "Sample."

3.13 Most samples have 5 to 10 fast-growing rats.

CHAPTER 4

4.1 Statistic.

4.3 (a) It is legitimate. (b) 35%.

4.5 Each possible value (1,2,3,4) has probability 1/4.

4.7 Possible totals: 2 through 8; probabilities 1/16, 2/16, 3/16, 4/16, 3/16, 2/16, 1/16.

4.9 (a) Properties of random numbers like these guarantee that each of the 10,000 distinct four-digit numbers (0000 through 9999) is equally likely. (b) 8888.

4.11 (b) A personal probability might take into account specific information about your driving habits. (c) Most people believe that they are better-than-average drivers.

4.13 (a) 19, 22, 39, 50, 34; $\hat{p} = 0.6$. (b) Two samples give $\hat{p} = 0.4$, four give 0.6, three give 0.8, and one gives 1. (d) Since an SRS is unbiased, 0.6 should be in the center.

4.15 About 0.0004.

4.17 Results will vary.

4.19 The histograms should be centered at about 0.6 and 0.2 (with quite a bit of spread).

4.21 The mean is 10.5.

CHAPTER 5

5.1 (a) 0.771. (b) 0.48.

5.3 4% of adults go to health clubs at least twice per week.

5.5 (a) 0.25. (b) The distribution is skewed to the right.

5.7 10%.

5.9 $P(D) = 0.4$.

CHAPTER 6

6.1 (a) Such results would rarely occur if vitamin C were ineffective. (b) Such results would occur less than 1% of the time if vitamin C were ineffective.

6.3 (a) No treatments were prescribed. (b) Any difference between cell phone users and non-users might arise from chance variation alone. (c) 5% significance tests give false positives 5% of the time.

6.5 (a) The differences observed might occur by chance even if SES had no effect. (b) This tells us that the test was not insignificant merely because of a small sample size.

6.7 Were these random samples? How big were the samples?

6.9 The increase in fires over time would occur less than 1% of the time by chance. This straight-line relationship explains 61% of the variation in the number of fires.

6.11 (a) Yes. (b) No. (c) No.

6.13 (a) Most answers: At least 40 of the 50, and between 87% and 93% of the 1000. (b) Most answers: At least 44 of the 50, and between 93% and 97% of the 1000. (c) Most answers: At least 47 of the 50, and at least 98% of the 1000.

6.17 ±2.24.

6.19 $z^* = 1.78$; 119.2 to 128.4 bushels/acre.

CHAPTER 7

7.1 (a) Use a paired-sample test. (b) $t = 10.16$, $P < 0.00005$. Coached students do improve their scores. (c) About 21.5 to 36.5 points.

7.3 This was an observational study, not an experiment.

7.5 (a) 22.6 to 26.9 degrees. (b) This interval is narrower.

7.7 2.624.

7.9 2.5 to 3.3 ng C per liter.

7.11 (a) Fund and index performances are certainly not independent. (c) $t = 0.36$ with 18 degrees of freedom; not significant.

7.13 (a) Standard error of the mean, which equals $s/\sqrt{n}$. (b) A two-sample t test. (c) The difference in insulin (glucose) levels would happen less than 5% (0.5%) of the time if the two populations were identical. $P < 0.005$ is stronger evidence.

CHAPTER 8

8.1 $z \doteq -2.40$, $P = 0.0164$—fairly strong evidence against H_0.

8.3 0.3577 to 0.4723.

8.5 H_0: $p_1 = p_2$; H_a: $p_1 < p_2$; $z \doteq -4.82$; P is tiny; we conclude that bupropion increases the success rate.

8.7 $z \doteq 1.175$, $P > 0.10$, we cannot conclude that more than half of Americans hold this opinion.

8.9 $z \doteq -2.25$; $P = 0.0122$; black parents are less likely to rate schools favorably.

8.11 Start with $\hat{p} \doteq 0.4047$ and a standard error of 0.01416.

CHAPTER 9

9.1 (a) Success rates: 0.1639, 0.3033, 0.3551, 0.15625. (b) To make the two-way table, number of failures equals subjects minus successes. To test the hypotheses (association versus no association), $X^2 = 34.937$ (3 df), $P < 0.0005$; the drug, or the patch/drug combination, seem to be effective.

9.3 Response rates: 84.4%, 94.5%, 100%. Percent responding appears to increase with level of vibration.

9.5 $X^2 = 22.426$ (8 df), $P = 0.004$. Blacks are less likely, and Hispanics more likely, to consider schools excellent, while Hispanics and whites differ in percentage considering schools good (whites are higher) and percentage who "don't know" (Hispanics are higher).

9.7 (a) Proportions in favor: 0.5, 0.3944, 0.3650, 0.4206, and 0.4375. (b) $X^2 = 8.525$ (4 df), $P = 0.075$; cannot conclude that the proportion varies with level of education.

CHAPTER 10

10.1 (a) Randomly allocate 36 flies to each of four groups, which receive varying dosages of caffeine; observe length of rest periods. (b) H_0: All groups have the same mean rest period; H_a: At least one group has a different mean rest period. We conclude that caffeine reduces the length of the rest period.

10.3 (a) Yes; the mean control emission rate is half the smallest of the others. (b) H_0: All groups have the same mean emission rate. H_a: At least one group has a different mean emission rate. (c) Are the data normally distributed? Were these random samples? (d) $s = \text{SEM} \times \sqrt{8}$. The rule-of-thumb ratio is 1.4755.

10.5 $\bar{x} \doteq 21.585$, $\text{MSG} \doteq 745.5$, $\text{MSE} \doteq 460.2$, degrees of freedom 3 and 28, $F \doteq 1.62$—not significant.

CHAPTER 11

11.1 (a) Large samples give significance for small relationships. (b) About 9×10^{-47}.

11.3 (a) H_0: $\beta = 0$; H_a: $\beta \neq 0$; $t = -1.27$ and $P = 0.226$; cannot reject H_0. (b) The scatterplot shows considerable variation about the regression line.

4.4 Control Charts

There are many situations in which our goal is to hold a variable constant over time. You may monitor your weight or blood pressure and plan to modify your behavior if either changes. Manufacturers watch the results of regular measurements made during production and plan to take action if quality deteriorates. Statistics plays a central role in these situations because of the presence of variation. *All processes have variation.* Your weight fluctuates from day to day; the critical dimension of a machined part varies a bit from item to item. Variation occurs in even the most precisely made product due to small changes in the raw material, the adjustment of the machine, the behavior of the operator, and even the temperature in the plant. Because variation is always present, we can't expect to hold a variable exactly constant over time. The statistical description of stability over time requires that the pattern of variation remain stable, not that there be no variation in the variable measured.

> **STATISTICAL CONTROL**
>
> A variable that continues to be described by the same distribution when observed over time is said to be in statistical control, or simply **in control**.
>
> **Control charts** are statistical tools that monitor the control of a process and alert us when the process has been disturbed.

Control charts work by distinguishing the natural variation in the process from the additional variation that suggests that the process has changed. A control chart sounds an alarm when it sees too much variation. The most common application of control charts is to monitor the performance of an industrial process. The same methods, however, can be used to check the stability of quantities as varied as the ratings of a television show, the level of ozone in the atmosphere, and the gas mileage of your car. Control charts combine graphical and numerical descriptions of data with use of sampling distributions. They therefore provide a natural bridge between exploratory data analysis and formal statistical inference.*

$\bar{x}$ charts

The population in the control chart setting is all items that would be produced by the process if it ran on forever in its present state. The items actually produced form samples from this population. We generally speak of the process rather than the population. Choose a quantitative variable, such as a diameter or a voltage, that is an important measure of the quality of an item. The process mean μ is the long-term average value of this variable; μ describes the

---

*Control charts were invented in the 1920s by Walter Shewhart at the Bell Telephone Laboratories. Shewhart's classic book, *Economic Control of Quality of Manufactured Product* (Van Nostrand, New York, 1931), organized the application of statistics to improving quality.

Table 4.1 $\bar{x}$ from 20 samples of size 4

Sample	1	2	3	4	5	6	7	8	9	10
$\bar{x}$	269.5	297.0	269.6	283.3	304.8	280.4	233.5	257.4	317.5	327.4
Sample	11	12	13	14	15	16	17	18	19	20
$\bar{x}$	264.7	307.7	310.0	343.3	328.1	342.6	338.8	340.1	374.6	336.1

center or aim of the process. The sample mean $\bar{x}$ of several items estimates μ and helps us judge whether the center of the process has moved away from its proper value. The most common control chart plots the means $\bar{x}$ of small samples taken from the process at regular intervals over time.

EXAMPLE 4.15 Making computer monitors

A manufacturer of computer monitors must control the tension on the mesh of fine wires that lies behind the surface of the viewing screen. Too much tension will tear the mesh, and too little will allow wrinkles. Tension is measured by an electrical device with output readings in millivolts (mV). The proper tension is 275 mV. Some variation is always present in the production process. When the process is operating properly, the standard deviation of the tension readings is $\sigma = 43$ mV.

The operator measures the tension on a sample of 4 monitors each hour. The mean $\bar{x}$ of each sample estimates the mean tension μ for the process at the time of the sample. Table 4.1 shows the observed $\bar{x}$'s for 20 consecutive hours of production. How can we use these data to keep the process in control?

A time plot helps us see whether or not the process is stable. Figure 4.13 is a plot of the successive sample means against the order in which the samples were taken. Because the target value for the process mean is $\mu = 275$ mV, we draw a *center line* at that level across the plot. The means from the later samples fall above this line and are consistently higher than those from earlier samples. This suggests that the process mean μ may have shifted upward, away from its target value of 275 mV. But perhaps the drift in $\bar{x}$ simply reflects the natural variation in the process. We need to back up our graph by calculation.

We expect $\bar{x}$ to have a distribution that is close to normal. Not only are the tension measurements roughly normal, but also the central limit theorem effect implies that sample means will be closer to normal than individual measurements. Because a control chart is a warning device, it is not necessary that our probability calculations be exactly correct. Approximate normality is good enough. In that same spirit, control charts use the approximate normal probabilities given by the 68–95–99.7 rule rather than more exact calculations using Table A.

If the standard deviation of the individual screens remains at $\sigma = 43$ mV, the standard deviation of $\bar{x}$ from 4 screens is

$$\frac{\sigma}{\sqrt{n}} = \frac{43}{\sqrt{4}} = 21.5 \text{ mV}$$

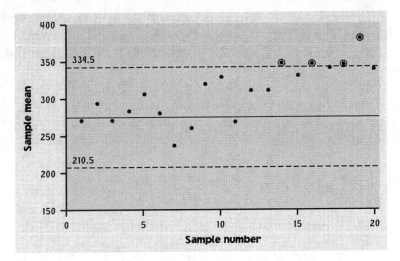

Figure 4.13 An $\bar{x}$ control chart for the data in Table 4.1. The points plotted are mean tension measurements $\bar{x}$ for samples of 4 computer monitor screens taken hourly during production. The center line and control limits help determine whether the process has been disturbed.

As long as the mean remains at its target value $\mu = 275$ mV, the 99.7 part of the 68–95–99.7 rule says that almost all values of $\bar{x}$ will lie between

$$\mu - 3\frac{\sigma}{\sqrt{n}} = 275 - (3)(21.5) = 210.5$$

$$\mu + 3\frac{\sigma}{\sqrt{n}} = 275 + (3)(21.5) = 339.5$$

We therefore draw dashed *control limits* at these two levels on the plot. We now have an $\bar{x}$ control chart.

> **$\bar{x}$ CONTROL CHART**
>
> To evaluate the control of a process with given standards μ and σ, make an $\bar{x}$ **control chart** as follows:
> - Plot the means $\bar{x}$ of regular samples of size n against time.
> - Draw a horizontal **center line** at μ.
> - Draw horizontal **control limits** at $\mu \pm 3\sigma/\sqrt{n}$.
>
> Any $\bar{x}$ that does not fall between the control limits is evidence that the process is out of control.

Four points, which are circled in Figure 4.13, lie above the upper control limit of the control chart. The 99.7 part of the 68–95–99.7 rule says that the probability is only 0.003 that a particular point would fall outside the control limits if μ and σ remain at their target values. These points are therefore good evidence that the distribution of mesh tension has changed. It appears that the process mean moved up. In practice, the operators search for a disturbance in the process as soon as they notice the first out-of-control point, that is, after sample number 14. Lack of control might be caused by a new operator, a new batch of mesh, or a breakdown in the tensioning apparatus. The out-of-control signal alerts us to the change before a large number of defective monitors are produced.

$\bar{x}$ chart

An $\bar{x}$ control chart is often called simply an $\bar{x}$ **chart**. Points $\bar{x}$ that vary between the control limits of an $\bar{x}$ chart represent the chance variation that is present in a normally operating process. Points that are out of control suggest that some source of additional variability has disturbed the stable operation of the process. Such a disturbance makes out-of-control points probable rather than unlikely. For example, if the process mean μ in Example 4.15 shifts from 275 mV to 339.5 mV (which is the value of the upper control limit), the probability that the next point falls above the upper control limit increases from about 0.0015 to 0.5.

APPLY YOUR KNOWLEDGE

4.66 Calibrating thermostats. A maker of auto air conditioners checks a sample of 4 thermostats from each hour's production. The thermostats are set at 75°F and then placed in a chamber where the temperature rises gradually. The tester records the temperature at which the thermostat turns on the air conditioner. The target for the process mean is $\mu = 75°$. Past experience indicates that the response temperature of properly adjusted thermostats varies with $\sigma = 0.5°$. The mean response temperature $\bar{x}$ for each hour's sample is plotted on an $\bar{x}$ control chart. Calculate the center line and control limits for this chart.

4.67 Milling slots. The width of a slot cut by a milling machine is important to the proper functioning of an aircraft hydraulic system. The manufacturer checks the control of the milling process by measuring a sample of 5 consecutive items during each hour's production. The mean slot width for each sample is plotted on an $\bar{x}$ control chart. The target width for the slot is $\mu = 0.8750$ inch. When properly adjusted, the milling machine should produce slots with mean width equal to the target value and standard deviation $\sigma = 0.0012$ inch. What center line and control limits should you draw on the $\bar{x}$ chart?

Statistical process control

The purpose of a control chart is not to ensure good quality by inspecting most of the items produced. Control charts focus on the manufacturing process itself rather than on the products. By checking the process at regular intervals, we can detect disturbances and correct them quickly. This is called **statistical process control**. Process control achieves high quality at a lower cost than inspecting all of the products. Small samples of 4 or 5 items are usually adequate for process control.

statistical process control

A process that is in control is stable over time, but stability alone does not guarantee good quality. The natural variation in the process may be so large that many of the products are unsatisfactory. Nonetheless, establishing control brings a number of advantages.

- In order to assess whether the process quality is satisfactory, we must observe the process operating in control free of breakdowns and other disturbances.
- A process in control is predictable. We can predict both the quantity and the quality of items produced.
- When a process is in control we can easily see the effects of attempts to improve the process, which are not hidden by the unpredictable variation that characterizes lack of statistical control.

A process in control is doing as well as it can in its present state. If the process is not capable of producing adequate quality even when undisturbed, we must make some major change in the process, such as installing new machines or retraining the operators.

Using control charts

The basis for the $\bar{x}$ chart is the sampling distribution of the sample mean $\bar{x}$. This assumes that the individual observations are a random sample from the population of interest. If the usual 4 or 5 items in a sample are an SRS from an hour's production, the population is all items produced that hour. It is more common, however, to regularly sample 4 or 5 consecutive items. In that case the population exists only in our minds. It contains all items that would be produced by the process as it was operating at the time of sampling. The control chart monitors the state of the process at regular intervals to see if a change has taken place.

Deciding how to sample is an important part of process control in practice. Walter Shewhart, the inventor of statistical process control, used the term **rational subgroup** in place of "sample." He wanted to emphasize that what the control chart monitors depends on the way we sample. Plotting $\bar{x}$ for an SRS from each hour's production will control hourly average outputs. There may be lots of up-and-down movement within each hour, but our control chart will not detect this if the hourly average output remains stable. An $\bar{x}$ chart based on regular samples of 4 or 5 consecutive items, on the other hand, will signal if the "instantaneous" average output changes over time. There is no one right way to sample—it depends on the nature of the process you are trying to keep stable.

rational subgroup

The basic signal for lack of control in an $\bar{x}$ chart is a single point beyond the control limits. In practice, however, other signals are used as well. In particular, a *run signal* is almost always combined with the basic one-point-out signal.

> **OUT-OF-CONTROL SIGNALS**
>
> The most common signals for lack of control in an $\bar{x}$ chart are:
> - One point falling outside the control limits.
> - A **run** of 9 points in a row on the same side of the center line.
>
> Begin a search for the cause as soon as a chart shows either signal.

Nine consecutive points on the same side of the center line are unlikely to occur unless the process aim has moved away from the target. Think of 9 straight heads or 9 straight tails in tossing a coin. The run signal often responds to a gradual drift in the process mean before the one-point-out signal, while the one-point-out signal often catches a sudden shift in the process mean more quickly. The two signals together make a good team. In the $\bar{x}$ chart of Figure 4.13, the run signal does not give an out-of-control signal until sample number 20. The one-point-out signal alerts us at sample 14.

APPLY YOUR KNOWLEDGE

4.68 **Forming tablets.** A pharmaceutical manufacturer forms tablets by compressing together the active ingredient and various fillers. The operators measure the hardness of a sample from each lot of tablets in order to control the compression process. The target values for the hardness are $\mu = 11.5$ and $\sigma = 0.2$. Table 4.2 gives three sets of data, each representing $\bar{x}$ for 20 successive samples of $n = 4$ tablets. One set remains in control at the target value. In a second set, the process mean μ shifts suddenly to a new value. In a third, the process mean drifts gradually.

(a) What are the center line and control limits for an $\bar{x}$ chart for hardness?

(b) Draw a separate $\bar{x}$ chart for each of the three data sets. Circle any points that are beyond the control limits. Also, check for runs of 9 points above or below the center line and mark the ninth point of any run as being out of control.

(c) Based on your work in (b) and the appearance of the control charts, which set of data comes from a process that is in control? In which case does the process mean shift suddenly and at about which sample do you think that the mean changed? Finally, in which case does the mean drift gradually?

The real world: $\bar{x}$ and s charts

In practice we rarely know the process mean μ and the process standard deviation σ. We must therefore base the center line and control limits for an $\bar{x}$

Table 4.2 Three sets of $\bar{x}$ from 20 samples of size 4

Sample	Data set A	Data set B	Data set C
1	11.602	11.627	11.495
2	11.547	11.613	11.475
3	11.312	11.493	11.465
4	11.449	11.602	11.497
5	11.401	11.360	11.573
6	11.608	11.374	11.563
7	11.471	11.592	11.321
8	11.453	11.458	11.533
9	11.446	11.552	11.486
10	11.522	11.463	11.502
11	11.664	11.383	11.534
12	11.823	11.715	11.624
13	11.629	11.485	11.629
14	11.602	11.509	11.575
15	11.756	11.429	11.730
16	11.707	11.477	11.680
17	11.612	11.570	11.729
18	11.628	11.623	11.704
19	11.603	11.472	12.052
20	11.816	11.531	11.905

chart on estimates of μ and σ from past samples from the process. This works simply only if the process was in control when the past samples were taken.

What is more, we know that even a basic description of a distribution requires a measure of spread as well as center. It isn't enough to control the aim or center of the process. We must also control its variability. In practice, an $\bar{x}$ chart is always accompanied by another control chart that monitors the short-term variability of the process. The most common choice is an s chart. As its name suggests, the s chart is a chart against time of the standard deviations of our samples.

These real-world complications don't change the basic logic of control charts, but they do make the details messier. Here is an overview:

1. All control charts use a center line at the mean of the statistic being plotted and control limits 3 standard deviations on either side of this mean. We do this even if we are plotting a statistic (such as the sample standard deviation s) that does not have a normal distribution.

2. Usually we must estimate the mean and standard deviation from past data. This complicates the recipes for center line and control limits. The complications are built into tables of **control chart constants**.

control chart constants

Table 4.3 Control chart constants

Sample size	2	3	4	5	6	7	8	9	10
A	2.659	1.954	1.628	1.427	1.287	1.182	1.099	1.032	0.975
B	2.267	1.568	1.266	1.089	0.970	0.882	0.815	0.761	0.716

$\bar{x}$ AND s CONTROL CHARTS

To evaluate the control of a process based on past samples from the process, calculate the mean $\bar{x}$ and the standard deviation s for each of the samples. Take $\bar{\bar{x}}$ to be the average of the $\bar{x}$'s and $\bar{s}$ to be the average of the s's.

The $\bar{x}$ **chart** is a plot of the $\bar{x}$'s against time with center line $\bar{\bar{x}}$ and control limits $\bar{\bar{x}} \pm A\bar{s}$.

The s **chart** is a plot of the s's against time with center line $\bar{s}$ and control limits $\bar{s} \pm B\bar{s}$.

The **control chart constants** A and B depend on the size of the samples. They appear in Table 4.3.

EXAMPLE 4.16 Surface roughness

The roughness of the surface of a metal part after reaming is important to the quality of the part. The machinist measures the roughness for samples of 5 consecutive parts at regular intervals. Table 4.4 gives the means $\bar{x}$ and the standard deviations s for the last 20 samples. (These data are reported in Stephen B. Vardeman and J. Marcus Jobe, *Statistical Quality Assurance Methods for Engineers*, Wiley, New York, 1999. This book is an excellent source for more information about statistical process control.)

Calculate from Table 4.4 that the means we need are

$$\bar{\bar{x}} = \frac{1}{20}(34.6 + 46.8 + \quad + 21.0) = 32.1$$

$$\bar{s} = \frac{1}{20}(3.4 + 8.8 + \quad + 1.0) = 3.76$$

Table 4.4 Roughness measurements on metal parts, 20 samples of 5 parts each

Sample	1	2	3	4	5	6	7	8	9	10
Mean	34.6	46.8	32.6	42.6	26.6	29.6	33.6	28.2	25.8	32.6
Std. dev.	3.4	8.8	4.6	2.7	2.4	0.9	6.0	2.5	3.2	7.5
Sample	11	12	13	14	15	16	17	18	19	20
Mean	34.0	34.8	36.2	27.4	27.2	32.8	31.0	33.8	30.8	21.0
Std. dev.	9.1	1.9	1.3	9.6	1.3	2.2	2.5	2.7	1.6	1.0

Source: Stephen B. Vardeman and J. Marcus Jobe, *Statistical Quality Assurance Methods for Engineers*, Wiley, New York, 1999.

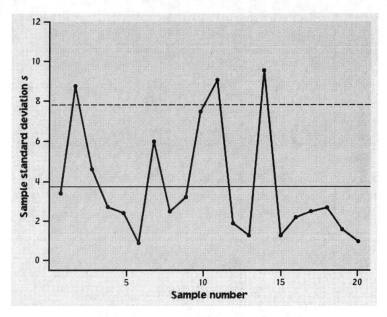

Figure 4.14 *An s control chart for the surface roughness data in Table 4.4. The control limits $\bar{s} \pm B\bar{s}$ are calculated from the data themselves.*

Always make the s chart first. Figure 4.14 plots s against the time order of the samples. The center line is at level $\bar{s} = 3.76$. The control limits use the control chart constant $B = 1.089$ for sample size 5 from Table 4.3. They are

$$\bar{s} \pm B\bar{s} = 3.76 \pm (1.089)(3.76)$$
$$= 3.76 \pm 4.09$$
$$= -0.33 \text{ and } 7.85$$

As often happens when the data show large variation, the lower control limit for the s chart is negative. Because s can never be negative, we ignore the lower limit and plot only the upper control limit in Figure 4.14.

The center line for the $\bar{x}$ chart is $\bar{\bar{x}} = 32.1$. The control limits are

$$\bar{\bar{x}} \pm A\bar{s} = 32.1 \pm (1.427)(3.76)$$
$$= 32.1 \pm 5.37$$
$$= 26.73 \text{ and } 37.47$$

The $\bar{x}$ chart appears in Figure 4.15.

To interpret $\bar{x}$ and s charts, proceed as follows. Look first at the s chart. It tells us if the variation in the process within the time represented by one sample stays stable. Figure 4.14 shows great variation in s from sample to sample. Three samples are out of control. If the s chart is not in control, stop and search for a cause. Perhaps a tool is loose in its holder, so that consecutive items sometimes but not always differ from each other.

The $\bar{x}$ chart tells us if the process remains stable over the longer periods of time that separate one sample from the next. Because $\bar{s}$ sets the control limits

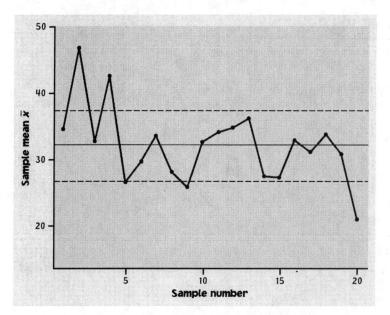

Figure 4.15 *An $\bar{x}$ control chart for the surface roughness data in Table 4.4. The control limits $\bar{x} \pm A\bar{s}$ are calculated from the data themselves.*

for the $\bar{x}$ chart, you cannot trust the $\bar{x}$ chart when s is out of control. A look at Figure 4.15 is nonetheless helpful. There seems to be a gradual downward trend in the mean roughness. Perhaps a tool is wearing, so that the mean roughness gradually decreases. Two early samples are out of control on the high side and the final sample is out of control low. If we find and fix the cause for the lack of control on the s chart, we expect future samples to have smaller $\bar{s}$. This will make the control limits for $\bar{x}$ tighter. The pattern of the $\bar{x}$ chart suggests some cause other than the one that is disturbing the s chart, so we expect more work after we bring s into control.

APPLY YOUR KNOWLEDGE

4.69 **Returns on a hot stock.** The rate of return on a stock varies from month to month. We can use a control chart to see if the pattern of variation is stable over time or whether there are periods during which the stock was unusually volatile by comparison with its own long-run pattern. We have data on the monthly returns (in percent) on Wal-Mart common stock during its early years of rapid growth, 1973 through 1991. Consider these 228 observations as 38 subgroups of 6 consecutive months each. Table 4.5 gives the means $\bar{x}$ and standard deviations s for these 38 samples of size $n = 6$.

Table 4.5 Wal-Mart stock performance, 38 6-month periods

Period	1	2	3	4	5	6	7	8	9	10
Mean	11.78	1.68	7.88	11.01	19.72	1.66	1.13	2.70	0.60	5.91
Std. dev.	14.69	31.53	19.28	7.13	27.50	9.44	8.56	12.64	10.02	4.79

Period	11	12	13	14	15	16	17	18	19	20
Mean	2.95	0.05	1.88	6.46	2.25	8.05	4.10	2.08	4.02	11.36
Std. dev.	10.87	9.80	2.57	15.72	10.68	4.99	6.49	5.90	7.87	6.50

Period	21	22	23	24	25	26	27	28	29	30
Mean	8.13	0.12	1.30	1.27	6.59	3.01	8.68	1.52	6.64	3.20
Std. dev.	8.61	5.93	8.57	5.02	8.13	9.32	7.04	7.72	6.43	15.08

Period	31	32	33	34	35	36	37	38
Mean	2.92	0.63	3.49	2.94	5.86	−0.25	6.02	5.87
Std. dev.	5.01	6.59	5.72	5.98	6.52	7.27	3.60	9.60

(a) Find the mean $\bar{\bar{x}}$ of the 38 $\bar{x}$'s and the mean $\bar{s}$ of the 38 s's.

(b) Use the control chart constant B from Table 4.3 to find the control limits for an s chart based on these past data. Make an s chart. Was the variation in the rate of return on Wal-Mart stock in a 6-month period generally about the same over the 19 years studied? Was there a trend in s over this period?

(c) Use the control chart constant A from Table 4.3 to find the control limits for an $\bar{x}$ chart based on these past data. Make an $\bar{x}$ chart. Based on your examination of the s chart, explain why the control limits for your $\bar{x}$ chart are so wide as to be of little use.

SECTION 4.4 Summary

A process that continues over time is **in control** if any variable measured on the process has the same distribution at any time. A process that is in control is operating under stable conditions.

An $\bar{x}$ **control chart** is a graph of sample means plotted against the time order of the samples, with a solid **center line** at the target value μ of the process mean and dashed **control limits** at $\mu \pm 3\sigma/\sqrt{n}$. An $\bar{x}$ chart helps us decide if a process is in control with mean μ and standard deviation σ.

The probability that the next point lies outside the control limits on an $\bar{x}$ chart is about 0.003 if the process is in control. Such a point is evidence that the process is **out of control**. That is, some disturbance has changed the distribution of the process. A cause for the change in the process should be sought.

There are other common signals for lack of control, such as a **run** of 9 consecutive points on the same side of the center line.

The purpose of **statistical process control** using control charts is to monitor a process so that any changes can be detected and corrected quickly. This is an economical method of maintaining good product quality.

In practice, control charts must be based on past data from the process. It is usual to keep both $\bar{x}$ **and** s **charts**. Control limits on these charts have the form "mean ± 3 standard deviations." For convenience, the details are built into tables of **control chart constants**.

Section 4.4 Exercises

4.70 Screw-on caps. A process molds plastic screw-on caps for containers of motor oil. The strength of properly made caps (the torque that would break the cap when it is screwed tight) has the normal distribution with mean 10 inch-pounds and standard deviation 1.2 inch-pounds. You monitor the molding process by testing a sample of 6 caps every 20 minutes. You measure the breaking strength of the sample caps and keep an $\bar{x}$ chart. Find the center line and control limits for this $\bar{x}$ chart.

4.71 More about screw-on caps. Suppose that you do not know the distribution of the strength of the screw-on caps in the previous exercise. But the process has been in control in the past and past samples of size 6 give $\bar{\bar{x}} = 10$ and $\bar{s} = 1.2$.

(a) Find the center line and control limits for an $\bar{x}$ chart from this information.

(b) The mean and standard deviation estimated from past data are the same as the mean and standard deviation given for the process in the previous exercise. The control limits based on past data are wider. Explain why we should expect this to happen.

4.72 Motor parts. The diameter of a bearing deflector in an electric motor is supposed to be 2.205 cm. Experience shows that when the manufacturing process is properly adjusted, it produces items with mean 2.2050 cm and standard deviation 0.0010 cm. A sample of 5 consecutive items is measured once each hour. The sample means and standard deviations for the past 12 hours are:

Hour	1	2	3	4	5	6
$\bar{x}$	2.2047	2.2047	2.2050	2.2049	2.2053	2.2043
s	0.0022	0.0012	0.0013	0.0005	0.0009	0.0004
Hour	7	8	9	10	11	12
$\bar{x}$	2.2036	2.2042	2.2038	2.2045	2.2026	2.2040
s	0.0011	0.0008	0.0008	0.0006	0.0008	0.0009

Make an $\bar{x}$ control chart for the deflector diameter using the given values of μ and σ. Use both the "one point out" and the "run of nine" signals to assess the control of the process. At what point should action have been taken to correct the process as the hourly point was added to the chart?

4.73 **More about motor parts.** In practice, we cannot be sure of the process μ and σ values given in the previous exercise. Use the data for the 12 past hours to estimate μ by $\bar{\bar{x}}$ and σ by $\bar{s}$. Make $\bar{x}$ and s control charts based on the past data. Is the s chart in control? What do you conclude from the $\bar{x}$ chart?

4.74 **Process control.** A manager who knows no statistics asks you, "What does it mean to say that a process is in control? Does being in control guarantee that the quality of our product is good?" Answer these questions in plain language that the manager can understand.

4.75 **Insulator strength.** Ceramic insulators are baked in lots in a large oven. After the baking, the process operator chooses 3 insulators at random from each lot, tests them for breaking strength, and plots the mean breaking strength for each sample on a control chart. The specifications call for a mean breaking strength of at least 10 pounds per square inch (psi). Past experience suggests that if the ceramic is properly formed and baked, the standard deviation in the breaking strength is about 1.2 psi. Here are the sample means and standard deviations from the last 15 lots:

Lot	1	2	3	4	5	6	7	8
$\bar{x}$	12.94	11.45	11.78	13.11	12.69	11.77	11.66	12.60
s	1.29	1.60	0.65	1.97	1.07	0.75	1.19	1.16

Lot	9	10	11	12	13	14	15
$\bar{x}$	11.23	12.02	10.93	12.38	7.59	13.17	12.14
s	0.92	1.27	1.33	0.84	0.60	1.65	0.46

(a) Make an s chart based on these data. Is the short-term variation of the process in control?

(b) Make an $\bar{x}$ chart based on the data. A process mean breaking strength greater than 10 psi is acceptable, so points out of control in the high direction do not call for remedial action. With this in mind, use both the one-point-out and run-of-nine signals to assess the control of the process and recommend action.

4.76 **Integrated circuits.** An important step in the manufacture of integrated circuit chips is etching the lines that conduct current between components on the chip. The chips contain a line width test pattern for process control measurements. The target width is

2.0 micrometers (μm). From long experience we know that the line width varies in production according to a normal distribution with mean $\mu = 1.829\mu$m and standard deviation $\sigma = 0.1516\mu$m.

(a) The acceptable range of line widths is $2.0 \pm 0.2\mu$m. What percent of chips have line widths outside this range?

(b) What are the control limits for an $\bar{x}$ chart for line width if samples of size 5 are measured at regular intervals during production? (Use the target value 2.0 as your center line.)

4.77 **Control limits.** The control limits of an $\bar{x}$ chart describe the normal range of variation of sample means. Don't confuse this with the range of values taken by individual items. Example 4.15 describes tension measurements for video screens, which vary with mean $\mu = 275$ and standard deviation $\sigma = 43$. The control limits, within which 99.7% of $\bar{x}$'s from samples of size 4 lie, are 210.5 and 339.5. A manager notices that many individual screens have tension outside this range. He is upset about this. Explain to the manager why many individuals fall outside the control limits. Then (assuming normality) find the range of values that contains 99.7% of individual tension measurements.

4.78 **Optional: European control charts.** The usual American and Japanese practice in making $\bar{x}$ charts is to place the control limits 3 standard deviations of $\bar{x}$ out from the center line. That is, the control limits are $\bar{x} \pm 3\sigma/\sqrt{n}$. The probability that a particular $\bar{x}$ falls outside these limits when the process is in control is about 0.003, using the 99.7 part of the 68–95–99.7 rule. European practice, on the other hand, places the control limits c standard deviations out, where the number c is chosen to give exactly probability 0.001 of a point $\bar{x}$ falling above $\mu + c\sigma/\sqrt{n}$ when the target μ and σ remain true. (The probability that $\bar{x}$ falls below $\mu - c\sigma/\sqrt{n}$ is also 0.001 because of the symmetry of the normal distributions.) Use Table A to find the value of c.

Sometimes we want to make a control chart for a single measurement x at each time period. **Control charts for individual measurements** are just $\bar{x}$ charts with the sample size $n = 1$. We cannot estimate the short-term process standard deviation σ from the individual samples because a sample of size 1 has no variation. Even with advanced methods of combining the information in several samples, the estimated σ will include some long-term variation and so will be too large. To compensate for this, it is common to use 2σ, rather than 3σ, control limits, that is, control limits $\mu \pm 2\sigma$. The next two exercises illustrate control charts for individual measurements. They also illustrate the use of control charts in personal record-keeping.

4.79 Optional: The professor swims. Professor Moore swims 2000 yards regularly. Here are his times (in minutes) for 23 sessions in the pool:

Time	34.12	35.72	34.72	34.05	34.13	35.72	36.17	35.57
Time	35.37	35.57	35.43	36.05	34.85	34.70	34.75	33.93
Time	34.60	34.00	34.35	35.62	35.68	35.28	35.97	

Find the mean $\bar{x}$ and standard deviation s for these times. Then make a control chart for the 23 times with center line at $\bar{x}$ and control limits at $\bar{x} \pm 2s$. Are the professor's swimming times in control? If not, describe the nature of any lack of control. (Because s measures variation among times over the full span of the 23 sessions in the pool, these control limits are wider than would be the case if we used advanced methods to estimate shorter-term variability.)

4.80 Optional: Drive time. Professor Moore records the time he takes to drive to the college each morning. Here are the times (in minutes) for 42 consecutive weekdays, with the dates in order along the rows:

8.25	7.83	8.30	8.42	8.50	8.67	8.17	9.00	9.00	8.17	7.92
9.00	8.50	9.00	7.75	7.92	8.00	8.08	8.42	8.75	8.08	9.75
8.33	7.83	7.92	8.58	7.83	8.42	7.75	7.42	6.75	7.42	8.50
8.67	10.17	8.75	8.58	8.67	9.17	9.08	8.83	8.67		

He also noted unusual occurrences on his record sheet: on October 27, a truck backing into a loading dock delayed him, and on December 5, ice on the windshield forced him to stop and clear the glass.

(a) Find $\bar{x}$ and s for the driving times.

(b) Plot the driving times against the order in which the observations were made. Add a center line and the control limits $\bar{x} \pm 2s$ to your chart. (The standard deviation s of all observations includes any long-term variation, so these limits are a bit crude.)

(c) Comment on the control of the process. Can you suggest explanations for individual points that are out of control? Is there any indication of an upward or downward trend in driving time?

Introduction

The most commonly used methods for inference about the means of quantitative response variables assume that the variables in question have normal distributions in the population or populations from which we draw our data. In practice, of course, no distribution is exactly normal. Fortunately, our usual methods for inference about population means (the one-sample and two-sample t procedures and analysis of variance) are quite **robust**. That is, the results of inference are not very sensitive to moderate lack of normality, especially when the samples are reasonably large. Some practical guidelines for taking advantage of the robustness of these methods appear in Chapter 7.

robustness

What can we do if plots suggest that the data are clearly not normal, especially when we have only a few observations? This is not a simple question. Here are the basic options:

outliers

1. If there are extreme **outliers** in a small data set, any inference method may be suspect. An outlier is an observation that may not come from the same population as the others. To decide what to do, you must find the cause of the outlier. Equipment failure that produced a bad measurement, for example, entitles you to remove the outlier and analyze the remaining data. If the outlier appears to be "real data," it is risky to draw any conclusion from just a few observations. This is the advice we gave to the child development researcher in Example 2.13.

transforming data

2. Sometimes we can **transform** our data so that their distribution is more nearly normal. Transformations, such as the logarithm, that pull in the long tail of right-skewed distributions are particularly helpful. We used the logarithm transformation in Exercises 2.15 and 2.96.

3. In some settings, **other standard distributions** replace the normal distributions as models for the overall pattern in the population. The lifetimes in service of equipment or the survival times of cancer patients after treatment usually have right-skewed distributions. Statistical studies in these areas use families of right-skewed distributions rather than normal distributions. There are inference procedures for the parameters of these distributions that replace the t procedures.

4. Finally, there are inference procedures that do not assume any specific form for the distribution of the population. These are called **nonparametric methods**. They are the subject of this chapter.

nonparametric methods

This chapter concerns one type of nonparametric procedure, tests that can replace the t tests and one-way analysis of variance when the normality assumptions for those tests are not met. The most useful nonparametric tests are **rank tests** based on the rank (place in order) of each observation in the set of all the data.

rank tests

Figure 12.1 presents an outline of the standard tests (based on normal distributions) and the rank tests that compete with them. All of these tests require that the population or populations have **continuous distributions**. That is, each distribution must be described by a density curve that allows observations to take any value in some interval of outcomes. The normal curves are one shape of density curve. Rank tests allow curves of any shape.

continuous distribution

Setting	Normal test	Rank test
One sample	One-sample t test Section 7.1	Wilcoxon signed rank test Section 12.2
Matched pairs	Apply one-sample test to differences within pairs	
Two independent samples	Two-sample t test Section 7.2	Wilcoxon rank sum test Section 12.1
Several independent samples	One-way ANOVA F test Chapter 10	Kruskal-Wallis test Section 12.3

Figure 12.1 *Comparison of tests based on normal distributions with nonparametric tests for similar settings.*

The rank tests we will study concern the *center* of a population or populations. When a population has at least roughly a normal distribution, we describe its center by the mean. The "normal tests" in Figure 12.1 all test hypotheses about population means. When distributions are strongly skewed, we often prefer the median to the mean as a measure of center. In simplest form, the hypotheses for rank tests just replace mean by median.

We devote a section of this chapter to each of the rank procedures. Section 12.1, which discusses the most common of these tests, also contains general information about rank tests. The kind of assumptions required, the nature of the hypotheses tested, the big idea of using ranks, and the contrast between exact distributions for use with small samples and approximations for use with larger samples are common to all rank tests. Sections 12.2 and 12.3 more briefly describe other rank tests.

12.1 The Wilcoxon Rank Sum Test

Two-sample problems (see Section 7.2) are among the most common in statistics. The most useful nonparametric significance test compares two distributions. Here is an example of this setting.

EXAMPLE 12.1   Weeds among the corn

Does the presence of small numbers of weeds reduce the yield of corn? Lamb's-quarter is a common weed in corn fields. A researcher planted corn at the same rate in 8 small plots of ground, then weeded the corn rows by hand to allow no weeds in 4 randomly selected plots and exactly 3 lamb's-quarter plants per meter of row in the other 4 plots. Here are the yields of corn (bushels per acre) in each of the plots. (Data provided by Samuel Phillips, Purdue University.)

Weeds per meter	Yield (bu./acre)			
0	166.7	172.2	165.0	176.9
3	158.6	176.4	153.1	156.0

Figure 12.2 Back-to-back stemplots of corn yields from plots with no weeds and with 3 weeds per meter of row. We split the stems, with leaves 0 to 4 on the first stem and leaves 5 to 9 on the second stem.

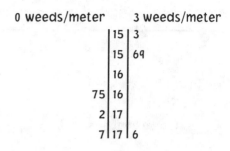

A back-to-back stemplot (Figure 12.2) suggests that yields may be lower when weeds are present. There is one outlier, which is correct data and cannot be removed. The samples are too small to rely on the robustness of the two-sample t test. We may prefer to use a test that does not require normality.

Ranks

We first rank all 8 observations together. To do this, arrange them in order from smallest to largest:

153.1   156.0   158.6   **165.0**   **166.7**   **172.2**   176.4   **176.9**

The boldface entries in the list are the yields with no weeds present. We see that four of the five highest yields come from that group, suggesting that yields are higher with no weeds. The idea of rank tests is to look just at position in this ordered list. To do this, replace each observation by its order, from 1 (smallest) to 8 (largest). These numbers are the *ranks*:

Yield	153.1	156.0	158.6	**165.0**	**166.7**	**172.2**	176.4	**176.9**
Rank	1	2	3	4	5	6	7	8

> **RANKS**
>
> To rank observations, first arrange them in order from smallest to largest. The **rank** of each observation is its position in this ordered list, starting with rank 1 for the smallest observation.

Moving from the original observations to their ranks retains only the ordering of the observations and makes no other use of their numerical values. Working with ranks allows us to dispense with specific assumptions about the shape of the distribution, such as normality.

The Wilcoxon rank sum test

If the presence of weeds reduces corn yields, we expect the ranks of the yields from plots with weeds to be smaller as a group than the ranks from plots without weeds. We might compare the *sums* of the ranks from the two treatments:

Treatment	Sum of ranks
No weeds	23
Weeds	13

These sums measure how much the ranks of the weed-free plots as a group exceed those of the weedy plots. In fact, the sum of the ranks from 1 to 8 is always equal to 36, so it is enough to report the sum for one of the two groups. If the sum of the ranks for the weed-free group is 23, the ranks for the other group must add to 13 because $23 + 13 = 36$. If the weeds have no effect, we would expect the sum of the ranks in either group to be 18 (half of 36). Here are the facts we need in a more general form that takes account of the fact that our two samples need not be the same size.

THE WILCOXON RANK SUM TEST

Draw an SRS of size n_1 from one population and draw an independent SRS of size n_2 from a second population. There are N observations in all, where $N = n_1 + n_2$. Rank all N observations. The sum W of the ranks for the first sample is the **Wilcoxon rank sum statistic**. If the two populations have the same continuous distribution, then W has mean

$$\mu_W = \frac{n_1(N+1)}{2}$$

and standard deviation

$$\sigma_W = \sqrt{\frac{n_1 n_2 (N+1)}{12}}$$

The **Wilcoxon rank sum test** rejects the hypothesis that the two populations have identical distributions when the rank sum W is far from its mean.*

In the corn yield study of Example 12.1, we want to test

H_0: no difference in distribution of yields

against the one-sided alternative

H_a: yields are systematically higher in weed-free plots

Our test statistic is the rank sum $W = 23$ for the weed-free plots.

---

*This test was invented by Frank Wilcoxon (1892–1965) in 1945. Wilcoxon was a chemist who met statistical problems in his work at the research laboratories of the American Cyanamid company.

EXAMPLE 12.2 Weeds among the corn

In Example 12.1, $n_1 = 4$, $n_2 = 4$, and there are $N = 8$ observations in all. The sum of ranks for the weed-free plots has mean

$$\mu_W = \frac{n_1(N+1)}{2}$$

$$= \frac{(4)(9)}{2} = 18$$

and standard deviation

$$\sigma_W = \sqrt{\frac{n_1 n_2 (N+1)}{12}}$$

$$= \sqrt{\frac{(4)(4)(9)}{12}} = \sqrt{12} = 3.464$$

Although the observed rank sum $W = 23$ is higher than the mean, it is only about 1.4 standard deviations high. We now suspect that the data do not give strong evidence that yields are higher in the population of weed-free corn.

The P-value for our one-sided alternative is $P(W \geq 23)$, the probability that W is at least as large as the value for our data when H_0 is true.

To calculate the P-value $P(W \geq 23)$, we need to know the sampling distribution of the rank sum W when the null hypothesis is true. This distribution depends on the two sample sizes n_1 and n_2. Tables are therefore a bit unwieldy, though you can find them in handbooks of statistical tables. Most statistical software will give you P-values, as well as carry out the ranking and calculate W. However, some software packages give only approximate P-values. You must learn what your software offers.

EXAMPLE 12.3 Using software

Figure 12.3 shows the output from a software package that calculates the exact sampling distribution of W. We see that the sum of the ranks in the weed-free group is $W = 23$, with P-value $P = 0.10$ against the one-sided alternative that weed-free plots have higher yields. There is some evidence that weeds reduce yield, considering that we have data from only four plots for each treatment. The evidence does not, however, reach the levels usually considered convincing.

Figure 12.3 *Output from the S-Plus statistical software for the data in Example 12.1. This program uses the exact distribution for W when the samples are small and there are no ties.*

```
            Exact Wilcoxon rank-sum test

data: 0weeds and 3weeds
rank-sum statistic W = 23, n = 4, m = 4, p-value= 0.100
alternative hypothesis: true mu is greater than 0
```

It is worth noting that the two-sample t test gives essentially the same result as the Wilcoxon test in Example 12.3 ($t = 1.554, P = 0.0937$). It is in fact somewhat unusual to find a strong disagreement between the conclusions reached by these two tests.

The normal approximation

The rank sum statistic W becomes approximately normal as the two sample sizes increase. We can then form yet another z statistic by standardizing W:

$$z = \frac{W - \mu_W}{\sigma_W}$$
$$= \frac{W - n_1(N+1)/2}{\sqrt{n_1 n_2 (N+1)/12}}$$

Use standard normal probability calculations to find P-values for this statistic. Because W takes only whole-number values, we use a trick called the **continuity correction** to improve the accuracy of the approximation. To apply the continuity correction when a variable takes only whole-number values, act as if each whole number occupies the entire interval from 0.5 below the number to 0.5 above it.

continuity correction

EXAMPLE 12.4 Using the normal approximation

The standardized rank sum statistic W in our corn yield example is

$$z = \frac{W - \mu_W}{\sigma_W} = \frac{23 - 18}{3.464} = 1.44$$

We expect W to be larger when the alternative hypothesis is true, so the approximate P-value is

$$P(Z \geq 1.44) = 0.0749$$

The continuity correction acts as if the whole number 23 occupies the entire interval from 22.5 to 23.5. We calculate the P-value $P(W \geq 23)$ as $P(W \geq 22.5)$ because the value 23 is included in the range whose probability we want. Here is the calculation:

$$P(W \geq 22.5) = P\left(\frac{W - \mu_W}{\sigma_W} \geq \frac{22.5 - 18}{3.464}\right)$$
$$= P(Z \geq 1.30)$$
$$= 0.0968$$

The continuity correction gives a result closer to the exact value $P = 0.10$.

We recommend always using either the exact distribution (from software or tables) or the continuity correction for the rank sum statistic W. The exact distribution is safer for small samples. As Example 12.4 illustrates, however, the normal approximation with the continuity correction is often adequate.

CHAPTER 12 • Nonparametric Tests

EXAMPLE 12.5 More software output

Mann-Whitney test

Figure 12.4 shows the output for our data from two more statistical programs. Minitab offers only the normal approximation, and it refers to the **Mann-Whitney test.** This is an alternate form of the Wilcoxon rank sum test. SAS carries out both the exact and approximate tests. SAS calls the rank sum S rather than W, and gives the mean 18 and standard deviation 3.464 as well as the z statistic 1.299 (using the continuity correction). SAS gives the approximate two-sided P-value as 0.1939, so the one-sided result is half this, $P = 0.0970$. This agrees with Minitab and (up to a small roundoff error) with our result in Example 12.4. This approximate P-value is close to the exact result $P = 0.1000$, given by SAS and in Figure 12.3.

```
Mann-Whitney Confidence Interval and Test

0 weeds     N = 4     Median =  169.45
3 weeds     N = 4     Median =  157.30
Point estimate for ETA1-ETA2 is 11.30
97.0 Percent C.I. for ETA1-ETA2 is (-11.40, 23.80)
W = 23.0
Test of ETA1 = ETA2 vs. ETA1 > ETA2 is significant
at 0.0970
```

(a) Minitab

```
        Wilcoxon Scores (Rank Sums) for Variable YIELD
                 Classified by Variable WEEDS

                 Sum of    Expected     Std Dev          Mean
WEEDS    N       Scores    Under H0     Under H0        Score

  0      4       23.0       18.0       3.46410162    5.75000000
  3      4       13.0       18.0       3.46410162    3.25000000

Wilcoxon 2-Sample Test       S = 23.0000

Exact P-Values
   (One-sided) Prob >= S               = 0.1000
   (Two-sided) Prob >= |S - Mean|      = 0.2000

Normal Approximation (with Continuity Correction of .5)
Z = 1.29904        Prob > |Z| = 0.1939
```

(b) SAS

Figure 12.4 *Output from the Minitab and SAS statistical software for the data in Example 12.1.* **(a)** *Minitab uses the normal approximation for the distribution of W.* **(b)** *SAS gives both the exact and approximate values.*

What hypotheses does Wilcoxon test?

Our null hypothesis is that weeds do not affect yield. Our alternative hypothesis is that yields are lower when weeds are present. If we are willing to assume that yields are normally distributed, or if we have reasonably large samples, we use the two-sample t test for means. Our hypotheses then become:

$$H_0: \mu_1 = \mu_2$$
$$H_a: \mu_1 > \mu_2$$

When the distributions may not be normal, we might restate the hypotheses in terms of population medians rather than means:

$$H_0: \text{median}_1 = \text{median}_2$$
$$H_a: \text{median}_1 > \text{median}_2$$

The Wilcoxon rank sum test provides a significance test for these hypotheses, but only if an additional assumption is met: both populations must have distributions of *the same shape*. That is, the density curve for corn yields with 3 weeds per meter looks exactly like that for no weeds except that it may slide to a different location on the scale of yields. The Minitab output in Figure 12.4a states the hypotheses in terms of population medians (which it calls "eta") and also gives a confidence interval for the difference between the two population medians.

The same-shape assumption is too strict to be reasonable in practice. Fortunately, the Wilcoxon test also applies in a much more general and more useful setting. It tests hypotheses that we can state in words as:

$$H_0: \text{two distributions are the same}$$
$$H_a: \text{one has values that are systematically larger}$$

A more exact statement of the "systematically larger" alternative hypothesis is a bit tricky, so we won't try to give it here. The hypotheses for rank tests really are "nonparametric" because they do not involve any specific parameter such as the mean or median. If the two distributions do have the same shape, the general hypotheses reduce to comparing medians. Many texts and computer outputs state the hypotheses in terms of medians, sometimes ignoring the same shape requirement. We recommend that you express the hypotheses in words rather than symbols. "Yields are systematically higher in weed-free plots" is easy to understand and is a good statement of the effect that the Wilcoxon test looks for.

Ties

The exact distribution for the Wilcoxon rank sum is obtained assuming that all observations in both samples take different values. This allows us to rank them all. In practice, however, we often find observations tied at the same value. What shall we do? The usual practice is to *assign all tied values the **average** of the ranks they occupy*. Here is an example with 6 observations:

average ranks

Observation	153	155	158	158	161	164
Rank	1	2	3.5	3.5	5	6

The tied observations occupy the third and fourth places in the ordered list, so they share rank 3.5.

The exact distribution for the Wilcoxon rank sum W only applies to data without ties. Moreover, the standard deviation σ_W must be adjusted if ties are present. The normal approximation can be used after the standard deviation is adjusted. Statistical software will detect ties, make the necessary adjustment, and switch to the normal approximation. In practice, software is required if you want to use rank tests when the data contain tied values.

It is sometimes useful to use rank tests on data that have very many ties because the scale of measurement has only a few values. Here is an example.

EXAMPLE 12.6 Food safety at fairs

Food sold at outdoor fairs and festivals may be less safe than food sold in restaurants because it is prepared in temporary locations and often by volunteer help. What do people who attend fairs think about the safety of the food served? One study asked this question of people at a number of fairs in the midwest:

How often do you think people become sick because of food they consume prepared at outdoor fairs and festivals?

The possible responses were:

1 = very rarely
2 = once in a while
3 = often
4 = more often than not
5 = always

In all, 303 people answered the question. Of these, 196 were women and 107 were men. Is there good evidence that men and women differ in their perceptions about food safety at fairs? (Data from Huey Chern Boo, "Consumers' perceptions and concerns about safety and healthfulness of food served at fairs and festivals," M.S. thesis, Purdue University, 1997.)

We should first ask if the subjects in Example 12.6 are a random sample of people who attend fairs, at least in the midwest. The researcher visited 11 different fairs. She stood near the entrance and stopped every 25th adult who passed. Because no personal choice was involved in choosing the subjects, we can reasonably treat the data as coming from a random sample. (As usual, there was some nonresponse, which could create bias.)

Here are the data, presented as a two-way table of counts:

| | Response | | | | | |
	1	2	3	4	5	Total
Female	13	108	50	23	2	196
Male	22	57	22	5	1	107
Total	35	165	72	28	3	303

Comparing row percentages shows that the women in the sample are more concerned about food safety then the men:

| | Response | | | | | |
	1	2	3	4	5	Total
Female	6.6%	55.1%	25.5%	11.7%	1.0%	100%
Male	20.6%	53.3%	20.6%	4.7%	1.0%	100%

Is the difference between the genders statistically significant?

We might apply the chi-square test (Chapter 9). It is highly significant ($X^2 = 16.120$, df $= 4, P = 0.0029$). Although the chi-square test answers our general question, it ignores the ordering of the responses and so does not use all of the available information. We would really like to know whether men or women are more concerned about the health of the food served. This question depends on the ordering of responses from least concerned to most concerned. We can use the Wilcoxon test for the hypotheses:

H_0: men and women do not differ in their responses
H_a: one of the two genders gives systematically larger responses than the other

The alternative hypothesis is two-sided. Because the responses can take only five values, there are very many ties. All 35 people who chose "very rarely" are tied at 1, and all 165 who chose "once in a while" are tied at 2.

EXAMPLE 12.7 Food safety: computer output

Figure 12.5 gives computer output for the Wilcoxon test. The rank sum for men (using average ranks for ties) is $W = 14{,}059.5$. The standardized value is $z = -3.33$ with two-sided P-value $P = 0.0009$. There is very strong evidence of a difference. Women are more concerned than men about the safety of food served at fairs.

Figure 12.5 *Output from SAS for the food safety study of Example 12.6. The approximate two-sided P-value is 0.0009.*

```
        Wilcoxon Scores (Rank Sums) for Variable SFAIR
                   Classified by Variable GENDER


                    Sum of    Expected    Std Dev        Mean
  GENDER    N       Scores    Under H0    Under H0       Score

  Female   196   31996.5000   29792.0    661.161398   163.247449
  Male     107   14059.5000   16264.0    661.161398   131.397196
                 Average Scores Were Used for Ties

  Wilcoxon 2-Sample Test (Normal Approximation)
  (with Continuity Correction of .5)

  S = 14059.5    Z = -3.33353    Prob > |Z| = 0.0009
```

With more than 100 observations in each group and no outliers, we might use the two-sample t even though responses take only five values. In fact, the results for Example 12.6 are $t = 3.3655$ with $P = 0.0009$. The P-value for two-sample t is the same as that for the Wilcoxon test. There is, however, another reason to prefer the rank test in this example. The t statistic treats the response values 1 through 5 as meaningful numbers. In particular, the possible responses are treated as though they are equally spaced. The difference between "very rarely" and "once in a while" is the same as the difference between "once in a while" and "often." This may not make sense. The rank test, on the other hand, uses only the order of the responses, not their actual values. The responses are arranged in order from least to most concerned about safety, so the rank test makes sense. Some statisticians avoid using t procedures when there is not a fully meaningful scale of measurement.

Limitations of nonparametric tests

The examples we have given illustrate the potential usefulness of nonparametric tests. Nonetheless, rank tests are of secondary importance relative to inference procedures based on the normal distribution.

- Nonparametric inference is largely restricted to simple settings. Normal inference extends to methods for use with complex experimental designs and multiple regression, but nonparametric tests do not. We stress normal inference in part because it leads on to more advanced statistics.

- Normal tests compare means, and are accompanied by simple confidence intervals for means or differences between means. When we use nonparametric tests to compare medians, we can also give confidence intervals, though they are rather awkward to calculate. However, the usefulness of nonparametric tests is clearest in settings when they do not simply compare medians—see the discussion of "What hypotheses does Wilcoxon test?" In these settings, there is no measure of the *size* of the observed effect that is closely related to the rank test of the *statistical significance* of the effect.

- The robustness of normal tests for means implies that we rarely encounter data that require nonparametric procedures to obtain reasonably accurate *P*-values. The *t* and *W* tests give very similar results in our examples. Nonetheless, many statisticians would not use a *t* test in Example 12.6 because the response variable gives only the order of the responses.

- There are more modern and more effective ways to escape the assumption of normality, based on new computational tools.

SECTION 12.1 Summary

Nonparametric tests do not require any specific form for the distribution of the population from which our samples come.

Rank tests are nonparametric tests based on the **ranks** of observations, their positions in the list ordered from smallest (rank 1) to largest. Tied observations receive the average of their ranks.

The **Wilcoxon rank sum test** compares two distributions to assess whether one has systematically larger values than the other. The Wilcoxon test is based on the **Wilcoxon rank sum statistic W**, which is the sum of the ranks of one of the samples. The Wilcoxon test can replace the **two-sample *t* test**.

***P*-values** for the Wilcoxon test are based on the sampling distribution of the rank sum statistic *W* when the null hypothesis (no difference in distributions) is true. You can find *P*-values from special tables, software, or a normal approximation (with continuity correction).

SECTION 12.1 Exercises

Statistical software is very helpful in doing these exercises. If you do not have access to software, base your work on the normal approximation with continuity correction.

12.1 **Tell me a story.** A study of early childhood education asked kindergarten students to tell a fairy tale that had been read to them earlier in the week. The 10 children in the study included 5 high progress readers and 5 low progress readers. Each child told two stories. Story 1 had been read to them; Story 2 had been read and also illustrated with pictures. An expert listened to a recording of the children and assigned a score for certain uses of language. Here are the data (provided by Susan Stadler, Purdue University):

Child	Progress	Story 1 score	Story 2 score
1	High	0.55	0.80
2	High	0.57	0.82
3	High	0.72	0.54
4	High	0.70	0.79
5	High	0.84	0.89
6	Low	0.40	0.77
7	Low	0.72	0.49
8	Low	0.00	0.66
9	Low	0.36	0.28
10	Low	0.55	0.38

Is there evidence that the scores of high progress readers are higher than those of low progress readers when they retell a story they have heard without pictures (Story 1)?

(a) Make back-to-back stemplots for the 5 responses in each group. Are any major deviations from normality apparent?

(b) Carry out a two-sample t test. State hypotheses and give the two sample means, the t statistic and its P-value, and your conclusion.

(c) Carry out the Wilcoxon rank sum test. State hypotheses and give the rank sum W for high progress readers, its P-value, and your conclusion. Do the t and Wilcoxon tests lead you to different conclusions?

12.2 Repeat the analysis of Exercise 12.1 for the scores when children retell a story they have heard and seen illustrated with pictures (Story 2).

12.3 **Wilcoxon step by step.** Use the data in Exercise 12.1 for children telling Story 2 to carry out by hand the steps in the Wilcoxon rank sum test.

(a) Arrange the 10 observations in order and assign ranks. There are no ties.

(b) Find the rank sum W for the 5 high progress readers. What are the mean and standard deviation of W under the null hypothesis that low progress and high progress readers do not differ?

(c) Standardize W to obtain a z statistic. Do a normal probability calculation with the continuity correction to obtain a one-sided P-value.

(d) The data for Story 1 contain tied observations. What ranks would you assign to the 10 scores for Story 1?

12.4 **Weeds among the corn.** The corn yield study of Example 12.1 also examined yields in four plots having 9 lamb's-quarter plants per meter of row. The yields (bushels per acre) in these plots were:

      162.8   142.4   162.7   162.4

There is a clear outlier, but re-checking the results found that this is the correct yield for this plot. The outlier makes us hesitant to use t procedures because $\bar{x}$ and s are not resistant.

(a) Is there evidence that 9 weeds per meter reduces corn yields when compared with weed-free corn? Use the Wilcoxon rank sum test with the data above and part of the data from Example 12.1 to answer this question.

(b) Compare the results from (a) with those from the two-sample t test for these data.

(c) Now remove the low outlier 142.4 from the data with 9 weeds per meter. Repeat both the Wilcoxon and t analyses. By how much did the outlier reduce the mean yield in its group? By how much did it increase the standard deviation? Did it have a practically important impact on your conclusions?

12.5 **DDT poisoning.** Exercise 7.37 reports the results of a study of the effect of the pesticide DDT on nerve activity in rats. The data for the DDT group are:

      12.207   16.869   25.050   22.429   8.456   20.589

The control group data are:

      11.074   9.686   12.064   9.351   8.182   6.642

It is difficult to assess normality from such small samples, so we might use a nonparametric test to assess whether DDT affects nerve response.

(a) State the hypotheses for the Wilcoxon test.

(b) Carry out the test. Report the rank sum W, its P-value, and your conclusion.

(c) The two-sample t test used in Exercise 7.37 found that $t = 2.9912$, $P = 0.0247$. Are your results different enough to change the conclusion of the study?

12.6 **Does polyester decay?** In Example 7.8, we compared the breaking strengths of polyester strips buried for 16 weeks with that of strips buried for 2 weeks. The breaking strengths in pounds were:

2 weeks	118	126	126	120	129
16 weeks	124	98	110	140	110

(a) Apply the Wilcoxon rank sum test to these data and compare your result with the $P = 0.189$ obtained from the two-sample t test in Example 7.8.

(b) What are the null and alternative hypotheses for the t test? For the Wilcoxon test?

12.7 **Each day I am getting better in math.** Table 7.6 (page 400) gives the pre-test and post-test scores for two groups of students taking a program to improve their basic mathematics skills. Did the treatment group show significantly greater improvement than the control group?

(a) Apply the Wilcoxon rank sum test to the post-test versus pre-test differences. Note that there are some ties. What do you conclude? Compare your findings with those from the two-sample t test in Exercise 7.33.

(b) What are the null and alternative hypotheses for each of the two tests we have applied to these data?

(c) What must we assume about the data to apply each of the tests?

12.8 **Logging in the rainforest.** Exercise 7.32 compared the number of tree species in unlogged plots in the rain forest of Borneo with the number of species in plots logged 8 years earlier. Here are the data:

Unlogged	22	18	22	20	15	21	13	13	19	13	19	15
Logged	17	4	18	14	18	15	15	10	12			

(a) Make a back-to-back stemplot of the data. Does there appear to be a difference in species counts for logged and unlogged plots?

(b) Does logging significantly reduce the mean number of species in a plot after 8 years? State the hypotheses, do a Wilcoxon test, and state your conclusion.

12.9 **Food safety in restaurants.** Example 12.6 describes a study of the attitudes of people attending outdoor fairs about the safety of the food served at such locations. The full data set is stored on the CD as the file eg12-06.dat. It contains the responses of 303 people to several questions. The variables in this data set are (in order):

subject   hfair   sfair   sfast   srest   gender

The variable "sfair" contains the responses described in the example concerning safety of food served at outdoor fairs and festivals. The variable "srest" contains responses to the same question asked about food served in restaurants. The variable "gender" contains 1 if the respondent is a woman, 2 if a man. We saw that women are more concerned than men about the safety of food served at fairs. Is this also true for restaurants?

12.10 More on food safety. The data file used in Example 12.6 and Exercise 12.9 contains 303 rows, one for each of the 303 respondents. Each row contains the responses of one person to several questions. We wonder if people are more concerned about safety of food served at fairs than they are about the safety of food served at restaurants. Explain carefully why we *cannot* answer this question by applying the Wilcoxon rank sum test to the variables "sfair" and "srest."

12.11 Shopping in secondhand stores. To study customers' attitudes toward secondhand stores, researchers interviewed samples of shoppers at two secondhand stores of the same chain in two cities. Here are data on the incomes of shoppers at the two stores, presented as a two-way table of counts. (From William D. Darley, "Store-choice behavior for pre-owned merchandise," *Journal of Business Research*, 27 (1993), pp. 17–31.)

Income code	Income	City 1	City 2
1	Under $10,000	70	62
2	$10,000 to $19,999	52	63
3	$20,000 to $24,999	69	50
4	$25,000 to $34,999	22	19
5	$35,000 or more	28	24

(a) Is there a relationship between city and income? Use the chi-square test to answer this question.

(b) The chi-square test ignores the ordering of the income categories. The data file ex12-11.dat contains data on the 459 shoppers in this study. The first variable is the city (City 1 or City 2) and the second is the income code as it appears in the table above (1 to 5). Is there good evidence that shoppers in one city have systematically higher incomes than in the other?

12.2 The Wilcoxon Signed Rank Test

We use the one-sample t procedures for inference about the mean of one population or for inference about the mean difference in a matched pairs setting. The matched pairs setting is more important because good studies are generally comparative. We will now meet a rank test for this setting.

EXAMPLE 12.8 Tell me a story

A study of early childhood education asked kindergarten students to tell a fairy tale that had been read to them earlier in the week. Each child told two stories. The first had been read to them and the second had been read and also illustrated with pictures. An expert listened to a recording of the children and assigned a score for certain uses of language. Here are the data for five "low progress" readers in a pilot study:

Child	1	2	3	4	5
Story 2	0.77	0.49	0.66	0.28	0.38
Story 1	0.40	0.72	0.00	0.36	0.55
Difference	0.37	0.23	0.66	0.08	0.17

(Data provided by Susan Stadler, Purdue University.) We wonder if illustrations improve how the children retell a story. We would like to test the hypotheses:

H_0: scores have the same distribution for both stories

H_a: scores are systematically higher for Story 2

Because this is a matched pairs design, we base our inference on the differences. The matched pairs t test gives $t = 0.635$ with one-sided P-value $P = 0.280$. We cannot assess normality from so few observations. We would therefore like to use a rank test.

Positive differences in Example 12.8 indicate that the children performed better telling Story 2. If scores are generally higher with illustrations, the positive differences should be farther from zero in the positive direction than the negative differences are in the negative direction. We therefore compare the **absolute values** of the differences, that is, their magnitudes without a sign. Here they are, with bold face indicating the positive values:

absolute value

0.37   0.23   **0.66**   0.08   0.17

Arrange these in increasing order and assign ranks, keeping track of which values were originally positive. Tied values receive the average of their ranks. If there are zero differences, discard them before ranking.

Absolute value	0.08	0.17	0.23	**0.37**	**0.66**
Rank	1	2	3	4	5

The test statistic is the sum of the ranks of the positive differences. (We could equally well use the sum of the ranks of the negative differences.) This is the *Wilcoxon signed rank statistic*. Its value here is $W^+ = 9$.

THE WILCOXON SIGNED RANK TEST FOR MATCHED PAIRS

Draw an SRS of size n from a population for a matched pairs study and take the differences in responses within pairs. Rank the absolute values of these differences. The sum W^+ of the ranks for the positive differences is the **Wilcoxon signed rank statistic**. If the distribution of the responses is not affected by the different treatments within pairs, then W^+ has mean

$$\mu_{W^+} = \frac{n(n+1)}{4}$$

and standard deviation

$$\sigma_{W^+} = \sqrt{\frac{n(n+1)(2n+1)}{24}}$$

The **Wilcoxon signed rank test** rejects the hypothesis that there are no systematic differences within pairs when the rank sum W^+ is far from its mean.

EXAMPLE 12.9 Tell me a story

In the story-telling study of Example 12.8, $n = 5$. If the null hypothesis (no systematic effect of illustrations) is true, the mean of the signed rank statistic is

$$\mu_{W^+} = \frac{n(n+1)}{4} = \frac{(5)(6)}{4} = 7.5$$

Our observed value $W^+ = 9$ is only slightly larger than this mean. The one-sided P-value is $P(W^+ \geq 9)$.

Figure 12.6 displays the output of two statistical programs. We see that the one-sided P-value for the Wilcoxon signed rank test with $n = 5$ observations and $W^+ = 9$ is $P = 0.4062$. This result differs from the t test result $P = 0.280$, but both tell us that this very small sample gives no evidence that seeing illustrations improves the story telling of low-progress readers.

The normal approximation

The distribution of the signed rank statistic when the null hypothesis (no difference) is true becomes approximately normal as the sample size becomes large. We can then use normal probability calculations (with the continuity correction) to obtain approximate P-values for W^+. Let's see how this works in the story-telling example, even though $n = 5$ is certainly not a large sample.

Figure 12.6 *Output from* (a) *S-Plus and* (b) *Data Desk for the story telling study of Example 12.9. These programs use the exact distribution of W^+ when the sample size is small and there are no ties.*

```
          Exact Wilcoxon signed-rank test

data: Story2-Story1

signed-rank statistic V = 9, n = 5, p-value = 0.4062

alternative hypothesis: true mu is greater than 0
```

(a) S-Plus

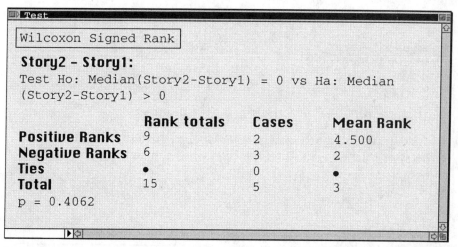

(b) Data Desk

EXAMPLE 12.10 Using the normal approximation

For $n = 5$ observations, we saw in Example 12.9 that $\mu_{W^+} = 7.5$. The standard deviation of W^+ under the null hypothesis is

$$\sigma_{W^+} = \sqrt{\frac{n(n+1)(2n+1)}{24}}$$

$$= \sqrt{\frac{(5)(6)(11)}{24}}$$

$$= \sqrt{13.75} = 3.708$$

The continuity correction calculates the P-value $P(W^+ \geq 9)$ as $P(W^+ \geq 8.5)$, treating the value $W^+ = 9$ as occupying the interval from 8.5 to 9.5. We find the normal approximation for the P-value by standardizing and using the standard normal table:

$$P(W^+ \geq 8.5) = P\left(\frac{W^+ - 7.5}{3.708} \geq \frac{8.5 - 7.5}{3.708}\right)$$

$$= P(Z \geq 0.27)$$

$$= 0.394$$

Despite the small sample size, the normal approximation gives a result quite close to the exact value $P = 0.4062$.

Ties

Ties among the absolute differences are handled by assigning average ranks. A tie *within* a pair creates a difference of zero. Because these are neither positive or negative, we drop such pairs from our sample. As in the case of the Wilcoxon rank sum, ties complicate finding a P-value. There is no longer a usable exact distribution for the signed rank statistic W^+, and the standard deviation σ_{W^+} must be adjusted for the ties before we can use the normal approximation. Software will do this. Here is an example.

EXAMPLE 12.11   Golf scores

Here are the golf scores of 12 members of a college women's golf team in two rounds of tournament play. (A golf score is the number of strokes required to complete the course, so that low scores are better.)

Player	1	2	3	4	5	6	7	8	9	10	11	12
Round 2	94	85	89	89	81	76	107	89	87	91	88	80
Round 1	89	90	87	95	86	81	102	105	83	88	91	79
Difference	5	5	2	6	5	5	5	16	4	3	3	1

Negative differences indicate better (lower) scores on the second round. We see that 6 of the 12 golfers improved their scores. We would like to test the hypotheses that in a large population of collegiate woman golfers

H_0: scores have the same distribution in rounds 1 and 2

H_a: scores are systematically lower or higher in round 2

A stemplot plot of the differences (Figure 12.7) shows some irregularity and a low outlier. We will use the Wilcoxon signed rank test.

The absolute values of the differences, with boldface indicating those that were negative, are:

5   5   2   6   5   5   5   **16**   4   3   **3**   1

```
-1 | 6
-1 |
-0 | 5 5 5 6
-0 | 3
 0 | 1 2 3 4
 0 | 5 5
```

Figure 12.7 *Stemplot of the differences in scores for two rounds of a golf tournament, from Example 12.11. We split the stems, with leaves 0 to 4 on the first stem and leaves 5 to 9 on the second stem.*

Arrange these in increasing order and assign ranks, keeping track of which values were originally negative. Tied values receive the average of their ranks.

Absolute value	1	2	3	3	4	5	5	5	5	5	6	16
Rank	1	2	3.5	3.5	5	8	8	8	8	8	11	12

The Wilcoxon signed rank statistic is the sum of the ranks of the negative differences. (We could equally well use the sum for the ranks of the positive differences.) Its value is $W^+ = 50.5$.

EXAMPLE 12.12 Golf scores: computer output

Here are the two-sided P-values for the Wilcoxon signed rank test for the golf score data from several statistical programs:

Program	P-value
Data Desk	$P = 0.366$
Minitab	$P = 0.388$
SAS	$P = 0.388$
S-PLUS	$P = 0.384$

All lead to the same practical conclusion: these data give no evidence for a systematic change in scores between rounds. However, the P-values reported differ a bit from program to program. The reason for the variations is that the programs use slightly different versions of the approximate calculations needed when ties are present. The exact result depends on which of these variations the programmer chooses to use.

For these data, the matched pairs t test gives $t = 0.9314$ with $P = 0.3716$. Once again, t and W^+ lead to the same conclusion.

SECTION 12.2 Summary

The **Wilcoxon signed rank test** applies to matched pairs studies. It tests the null hypothesis that there is no systematic difference within pairs against alternatives that assert a systematic difference (either one-sided or two-sided).

The test is based on the **Wilcoxon signed rank statistic W^+**, which is the sum of the ranks of the positive (or negative) differences when we rank the absolute values of the differences. The **matched pairs t test** and the **sign test** are alternative tests in this setting.

P-values for the signed rank test are based on the sampling distribution of W^+ when the null hypothesis is true. You can find P-values from special tables, software, or a normal approximation (with continuity correction).

Section 12.2 Exercises

Statistical software is very helpful in doing these exercises. If you do not have access to software, base your work on the normal approximation with continuity correction.

12.12 **Stepping up your heart rate (EESEE).** A student project asked subjects to step up and down for three minutes and measured their heart rates before and after the exercise. Here are data for five subjects and two treatments: stepping at a low rate (14 steps per minute) and at a medium rate (21 steps per minute). For each subject, we give the resting heart rate (beats per minutes) and the heart rate at the end of the exercise. (Simplified from the EESEE story "Stepping Up Your Heart Rate," on the CD.)

	Low Rate		Medium Rate	
Subject	Resting	Final	Resting	Final
1	60	75	63	84
2	90	99	69	93
3	87	93	81	96
4	78	87	75	90
5	84	84	90	108

Does exercise at the low rate raise heart rate significantly? State hypotheses in terms of the median increase in heart rate and apply the Wilcoxon signed rank test. What do you conclude?

12.13 **Stepping up your heart rate (EESEE).** Do the data from the previous exercise give good reason to think that stepping at the medium rate increases heart rates more than stepping at the low rate?

(a) State hypotheses in terms of comparing the median increases for the two treatments. What is the proper rank test for these hypotheses?

(b) Carry out your test and state a conclusion.

12.14 **Wilcoxon step by step.** Show the assignment of ranks and the calculation of the signed rank statistic W^+ for the data in Exercise 12.12. Remember that zeros are dropped from the data before ranking, so that n is the number of non-zero differences within pairs.

12.15 **Food safety at fairs and restaurants.** Example 12.6 describes a study of the attitudes of people attending outdoor fairs about the safety of the food served at such locations. The full data set is stored on the CD as the file eg12.06.dat. It contains the responses of 303 people to several questions. The variables in this data set are (in order):

subject   hfair   sfair   sfast   srest   gender

The variable "sfair" contains responses to the safety question described in Example 12.6. The variable "srest" contains responses to the same question asked about food served in restaurants. We suspect that restaurant food will appear safer than food served outdoors at a fair. Do the data give good evidence for this suspicion? (Give descriptive measures, a test statistic and its P-value, and your conclusion.)

12.16 Food safety at fairs and fast food restaurants. The food safety survey data described in Example 12.6 and Exercise 12.15 also contain the responses of the 303 subjects to the same question asked about food served at fast food restaurants. These responses are the values of the variable "sfast." Is there a systematic difference between the level of concern about food safety at outdoor fairs and at fast food restaurants?

12.17 Does nature heal best? Table 7.4 (page 387) gives data on the healing rate (micrometers per hour) of the skin of newts under two conditions. This is a matched pairs design, with the body's natural electric field for one limb (control) and half the natural value for another limb of the same newt (experimental). We want to know if the healing rates are systematically different under the two conditions. You decide to use a rank test.

State hypotheses, carry out a test, and give your conclusion. Be sure to include a description of what the data show in addition to the test results.

12.18 Sweetening colas. Cola makers test new recipes for loss of sweetness during storage. Trained tasters rate the sweetness before and after storage. Here are the sweetness losses (sweetness before storage minus sweetness after storage) found by 10 tasters for one new cola recipe:

2.0   0.4   0.7   2.0   0.4   2.2   1.3   1.2   1.1   2.3

Are these data good evidence that the cola lost sweetness?

(a) These data are the differences from a matched pairs design. State hypotheses in terms of the median difference in the population of all tasters, carry out a test, and give your conclusion.

(b) In Example 7.2 we found that the one-sample t test had P-value $P = 0.0122$ for these data. How does this compare with your result from (a)? What are the hypotheses for the *t* test? What assumptions must we make for each of the *t* and Wilcoxon tests?

12.19 **Ancient air.** Exercise 7.65 reports the following data on the percent of nitrogen in bubbles of ancient air trapped in amber:

63.4 65.0 64.4 63.3 54.8 64.5 60.8 49.1 51.0

We wonder if ancient air differs significantly from the present atmosphere, which is 78.1% nitrogen.

(a) Graph the data, and comment on skewness and outliers. A rank test is appropriate.

(b) We would like to test hypotheses about the median percent of nitrogen in ancient air (the population):

H_0: median = 78.1

H_a: median ≠ 78.1

To do this, apply the Wilcoxon signed rank statistic to the differences between the observations and 78.1. (This is the one-sample version of the test.) What do you conclude?

12.20 **Right versus left.** Table 7.2 (page 379) contains data from a student project that investigated whether right-handed people can turn a knob faster clockwise than they can counterclockwise. Describe what the data show, then state hypotheses and do a test that does not require normality. Report your conclusions carefully.

12.3 The Kruskal-Wallis Test

We have now considered alternatives to the paired-sample and two-sample *t* tests for comparing the magnitude of responses to two treatments. To compare more than two treatments, we use one-way analysis of variance (ANOVA) if the distributions of the responses to each treatment are at least roughly normal and have similar spreads. What can we do when these distribution requirements are violated?

EXAMPLE 12.13 Weeds among the corn

Lamb's-quarter is a common weed that interferes with the growth of corn. A researcher planted corn at the same rate in 16 small plots of ground, then randomly assigned the plots to four groups. He weeded the plots by hand to allow a fixed number of lamb's-quarter plants to grow in each meter of corn row. These numbers were 0, 1, 3, and 9 in the four groups of plots. No other weeds were allowed to grow, and all plots received identical treatment except for the weeds. Here are the yields of corn (bushels per acre) in each of the plots:

Weeds per meter	Corn yield	Weeds per meter	Corn yield	Weeds per meter	Corn yield	Weeds per meter	Corn yield
0	166.7	1	166.2	3	158.6	9	162.8
0	172.2	1	157.3	3	176.4	9	142.4
0	165.0	1	166.7	3	153.1	9	162.7
0	176.9	1	161.1	3	156.0	9	162.4

(Data from Samuel Phillips, Purdue University.) The summary statistics are:

Weeds	n	Mean	Std. dev.
0	4	170.200	5.422
1	4	162.825	4.469
3	4	161.025	10.493
9	4	157.575	10.118

The sample standard deviations do not satisfy our rule of thumb that for safe use of ANOVA the largest should not exceed twice the smallest. Moreover, we see that outliers are present in the yields for 3 and 9 weeds per meter. These are the correct yields for their plots, so we have no justification for removing them. We may want to use a nonparametric test.

Hypotheses and assumptions

The ANOVA F test concerns the means of the several populations represented by our samples. For Example 12.13, the ANOVA hypotheses are

$$H_0: \mu_0 = \mu_1 = \mu_3 = \mu_9$$
$$H_a: \text{not all four means are equal}$$

For example, μ_0 is the mean yield in the population of all corn planted under the conditions of the experiment with no weeds present. The data should consist of four independent random samples from the four populations, all normally distributed with the same standard deviation.

The *Kruskal-Wallis test* is a rank test that can replace the ANOVA F test. The assumption about data production (independent random samples from each population) remains important, but we can relax the normality assumption. We assume only that the response has a continuous distribution in each population. The hypotheses tested in our example are:

H_0: yields have the same distribution in all groups
H_a: yields are systematically higher in some groups than in others

If all of the population distributions have the same shape (normal or not), these hypotheses take a simpler form. The null hypothesis is that all four populations have the same *median* yield. The alternative hypothesis is that not all four median yields are equal.

The Kruskal-Wallis test

Recall the analysis of variance idea: we write the total observed variation in the responses as the sum of two parts, one measuring variation among the groups (sum of squares for groups, SSG) and one measuring variation among individual observations within the same group (sum of squares for error, SSE). The ANOVA F test rejects the null hypothesis that the mean responses are equal in all groups if SSG is large relative to SSE.

The idea of the Kruskal-Wallis rank test is to rank all the responses from all groups together, and then apply one-way ANOVA to the ranks rather than to the original observations. If there are N observations in all, the ranks are always the whole numbers from 1 to N. The total sum of squares for the ranks is therefore a fixed number no matter what the data are. So we do not need to look at both SSG and SSE. Although it isn't obvious without some unpleasant algebra, the Kruskal-Wallis test statistic is essentially just SSG for the ranks. We give the formula, but you should rely on software to do the arithmetic. When SSG is large, that is evidence that the groups differ.

> **THE KRUSKAL-WALLIS TEST**
>
> Draw independent SRSs of sizes $n_1, n_2, \ldots, n_I$ from I populations. There are N observations in all. Rank all N observations and let R_i be the sum of the ranks for the ith sample. The **Kruskal-Wallis statistic** is
>
> $$H = \frac{12}{N(N+1)} \sum \frac{R_i^2}{n_i} - 3(N+1)$$
>
> When the sample sizes n_i are large and all I populations have the same continuous distribution, H has approximately the chi-square distribution with $I - 1$ degrees of freedom.
>
> The **Kruskal-Wallis test** rejects the null hypothesis that all populations have the same distribution when H is large.

We now see that, like the Wilcoxon rank sum statistic, the Kruskal-Wallis statistic is based on the sums of the ranks for the groups we are comparing. The more different these sums are, the stronger is the evidence that responses are systematically larger in some groups than in others.

The exact distribution of the Kruskal-Wallis statistic H under the null hypothesis depends on all the sample sizes n_1 to n_I, so tables are awkward. The calculation of the exact distribution is so time-consuming for all but the smallest problems that even most statistical software uses the chi-square approximation to obtain P-values. As usual, there is no usable exact distribution when there are ties among the responses. We again assign average ranks to tied observations.

EXAMPLE 12.14 Weeds among the corn

In Example 12.13, there are $I = 4$ populations and $N = 16$ observations. The sample sizes are equal, $n_i = 4$. The 16 observations arranged in increasing order, with their ranks, are:

Yield	142.4	153.1	156.0	157.3	158.6	161.1	162.4	162.7
Rank	1	2	3	4	5	6	7	8
Yield	162.8	165.0	166.2	166.7	166.7	172.2	176.4	176.9
Rank	9	10	11	12.5	12.5	14	15	16

There is one pair of tied observations. The ranks for each of the four treatments are:

Weeds		Ranks			Sum of ranks
0	10	12.5	14	16	52.5
1	4	6	11	12.5	33.5
3	2	3	5	15	25.0
9	1	7	8	9	25.0

The Kruskal-Wallis statistic is therefore

$$H = \frac{12}{N(N+1)} \sum \frac{R_i^2}{n_i} - 3(N+1)$$

$$= \frac{12}{(16)(17)} \left(\frac{52.5^2}{4} + \frac{33.5^2}{4} + \frac{25^2}{4} + \frac{25^2}{4} \right) - (3)(17)$$

$$= \frac{12}{272}(1282.125) - 51$$

$$= 5.56$$

Referring to the table of chi-square critical points (Table E) with df = 3, we find that the P-value lies in the interval $0.10 < P < 0.15$. This small experiment suggests that more weeds decrease yield but does not provide convincing evidence that weeds have an effect.

```
Wilcoxon Scores (Rank Sums) for Variable YIELD
          Classified by Variable WEEDS

              Sum of     Expected      Std Dev        Mean
WEEDS   N     Scores     Under H0      Under H0       Score

0       4    52.5000000    34.0       8.24014563   13.1250000
1       4    33.5000000    34.0       8.24014563    8.3750000
3       4    25.0000000    34.0       8.24014563    6.2500000
9       4    25.0000000    34.0       8.24014563    6.2500000
             Average Scores Were Used for Ties

Kruskal-Wallis Test (Chi-Square Approximation)
CHISQ = 5.5725  DF = 3  Prob > CHISQ = 0.1344
```

Figure 12.8 *Output from SAS for the Kruskal-Wallis test applied to the data in Example 12.13. SAS uses the chi-square approximation to obtain a P-value.*

Figure 12.8 displays the output from the SAS statistical software, which gives the results $H = 5.5725$ and $P = 0.1344$. The software makes a small adjustment for the presence of ties that accounts for the slightly larger value of H. The adjustment makes the chi-square approximation more accurate. It would be important if there were many ties.

As an option, SAS will calculate the exact P-value for the Kruskal-Wallis test. The result for Example 12.14 is $P = 0.1299$. This result required several hours of computing time. Fortunately, the chi-square approximation is quite accurate. The ordinary ANOVA F test gives $F = 1.73$ with $P = 0.2130$. Although the practical conclusion is the same, ANOVA and Kruskal-Wallis do not agree closely in this example. The rank test is more reliable for these small samples with outliers.

Section 12.3 Summary

The **Kruskal-Wallis test** compares several populations on the basis of independent random samples from each population. This is the **one-way analysis of variance** setting.

The null hypothesis for the Kruskal-Wallis test is that the distribution of the response variable is the same in all the populations. The alternative hypothesis is that responses are systematically larger in some populations than in others.

The **Kruskal-Wallis statistic** H can be viewed in two ways. It is essentially the result of applying one-way ANOVA to the ranks of the observations. It is also a comparison of the sums of the ranks for the several samples.

When the sample sizes are not too small and the null hypothesis is true, H for comparing I populations has approximately the chi-square distribution with $I - 1$ degrees of freedom. We use this approximate distribution to obtain P-values.

Section 12.3 Exercises

Statistical software is needed to do these exercises without unpleasant hand calculations. If you do not have access to software, find the Kruskal-Wallis statistic H by hand and use the chi-square table to get approximate P-values.

12.21 How do nematodes (microscopic worms) affect plant growth? A botanist prepares 16 identical planting pots and then introduces different numbers of nematodes into the pots. A tomato seedling is transplanted into each plot. Here are data on the increase in height of the seedlings (in centimeters) 16 days after planting. (Data provided by Matthew Moore.)

Nematodes	Seedling growth			
0	10.8	9.1	13.5	9.2
1000	11.1	11.1	8.2	11.3
5000	5.4	4.6	7.4	5.0
10,000	5.8	5.3	3.2	7.5

We applied ANOVA to these data in Exercise 10.17. Because the samples are very small, it is difficult to assess normality.

(a) What hypotheses does ANOVA test? What hypotheses does Kruskal-Wallis test?

(b) Find the median growth in each group. Do nematodes appear to retard growth? Apply the Kruskal-Wallis test. What do you conclude?

12.22 Which color attracts beetles best? Example 10.6 used ANOVA to analyze the results of a study to see which of four colors best attracts cereal leaf beetles. Here are the data:

Color	Insects trapped					
Lemon yellow	45	59	48	46	38	47
White	21	12	14	17	13	17
Green	37	32	15	25	39	41
Blue	16	11	20	21	14	7

Because the samples are small, we will apply a nonparametric test.

(a) What hypotheses does ANOVA test? What hypotheses does Kruskal-Wallis test?

(b) Find the median number of beetles trapped by boards of each color. Which colors appear more effective? Use the Kruskal-Wallis

Table 12.1 Calories and sodium in three types of hot dogs

Beef hot dogs		Meat hot dogs		Poultry hot dogs	
Calories	Sodium	Calories	Sodium	Calories	Sodium
186	495	173	458	129	430
181	477	191	506	132	375
176	425	182	473	102	396
149	322	190	545	106	383
184	482	172	496	94	387
190	587	147	360	102	542
158	370	146	387	87	359
139	322	139	386	99	357
175	479	175	507	170	528
148	375	136	393	113	513
152	330	179	405	135	426
111	300	153	372	142	513
141	386	107	144	86	358
153	401	195	511	143	581
190	645	135	405	152	588
157	440	140	428	146	522
131	317	138	339	144	545
149	319				
135	298				
132	253				

Source: *Consumer Reports,* June 1986, pp. 366–367.

test to see if there are significant differences among the colors. What do you conclude?

12.23 How many calories in a hot dog? Table 12.1 presents data on the calorie and sodium content of selected brands of beef, meat, and poultry hot dogs. We will regard these brands as random samples from all brands available in food stores.

(a) Make stemplots of the calorie contents side by side, using the same stems for easy comparison. Give the five-number summaries for the three types of hot dog. What do the data suggest about the calorie content of different types of hot dog?

(b) Are any of the three distributions clearly not normal? Which ones, and why?

(c) Apply the Kruskal-Wallis test. Report your conclusions carefully.

12.24 Kruskal-Wallis step by step. Exercise 12.22 gives data on the counts of insects attracted by boards of four different colors. Carry out the Kruskal-Wallis test by hand, following these steps.

(a) What are I, the n_i, and N in this example?

(b) Arrange the counts in order and assign ranks. Be careful about ties. Find the sum of the ranks R_i for each color.

(c) Calculate the Kruskal-Wallis statistic H. How many degrees of freedom should you use for the chi-square approximation to its null distribution? Use the chi-square table to give an approximate P-value.

12.25 How much salt in a hot dog? Repeat the analysis of Exercise 12.23 for the sodium content of hot dogs.

12.26 Does polyester decay? Here are the breaking strengths (in pounds) of strips of polyester fabric buried in the ground for several lengths of time:

2 weeks	118	126	126	120	129
4 weeks	130	120	114	126	128
8 weeks	122	136	128	146	140
16 weeks	124	98	110	140	110

Breaking strength is a good measure of the extent to which the fabric has decayed.

(a) Find the standard deviations of the 4 samples. They do not meet our rule of thumb for applying ANOVA. In addition, the sample buried for 16 weeks contains an outlier. We will use the Kruskal-Wallis test.

(b) Find the medians of the four samples. What are the hypotheses for the Kruskal-Wallis test, expressed in terms of medians?

(c) Carry out the test and report your conclusion.

12.27 Food safety. Example 12.6 describes a study of the attitudes of people attending outdoor fairs about the safety of the food served at such locations. The full data set is stored on the CD as the file eg12-06.dat. It contains the responses of 303 people to several questions. The variables in this data set are (in order):

subject   hfair   sfair   sfast   srest   gender

The variable "sfair" contains responses to the safety question described in Example 12.6. The variables "srest" and "sfast" contain responses to the same question asked about food served in restaurants and in fast food chains. Explain carefully why we *cannot* use the Kruskal-Wallis test to see if there are systematic differences in perceptions of food safety in these three locations.

12.28 Logging in the rainforest. Table 10.3 (page 525) contains data comparing the number of trees and number of tree species in plots of land in a tropical rainforest that had never been logged with similar plots nearby that had been logged 1 year earlier and 8 years earlier.

(a) Use side by side stemplots to compare the distributions of number of trees per plot for the three groups of plots. Are there features that might prevent use of ANOVA? Also give the median number of trees per plot in the three groups.

(b) Use the Kruskal-Wallis test to compare the distributions of tree counts. State hypotheses, the test statistic and its P-value, and your conclusions.